Prespacetime Journal | May 2020 | Volume 11 | Issue 3

Prespacetime Journal

Volume 11 Issue 3
May 2020

Self-Organized Criticality, FTL in
Prespacetime, & Detection of Graviton

Editor:
Huping Hu, Ph.D., J.D.

Editors-at-Large:
Philip E. Gibbs, Ph.D.
Dainis Zeps, Ph.D.

Advisory Board

ISSN: 2153-8301   Prespacetime Journal   www.prespacetime.com
Published by QuantumDream, Inc.

Table of Contents

Articles

Derivation of the Muon *g*-2 Anomaly from Non-Equilibrium Dynamics
Ervin Goldfain
01-05

Solving the Flatness & Horizon Problems via Self-Organized Criticality
Ervin Goldfain
06-10

Quantum Field Theory as Manifestation of Self-Organized Criticality
Ervin Goldfain
11-19

Emergence of Lagrangian Field Theory from Self-Organized Criticality
Ervin Goldfain
20-29

Faster-Than-Light Anomalies in Prespacetime & Interstellar Translocation
through Hyperdimension
Chris H. Hardy
30-52

Math-Phys Article

Linear Codes over the Family of Finite Rings A_t
Abdullah Dertli & Yasemin Cengellenmis
53-60

Review Article

Detecting Gravitons on Black Hole Coalesence
Lawrence B. Crowell
61-66

Essays

Painting, Baking & Non-Associative Algebra
Jonathan J. Dickau
67-71

Is Gravity Curvature of Space-time?
Emre Dil
72-75

ISSN: 2153-8301
Prespacetime Journal
Published by QuantumDream, Inc.
www.prespacetime.com

(Published in Prespacetime Journal | May 2020 | Volume 11| Issue 3 | pp. 198-293)
Table of Contents

ii

Exploration

Derivation of the θ-parameter in Quantum Chromodynamics
Ervin Goldfain   76-78

Perspective

Towards Gross-Pitaevskiian Description of Solar System & Galaxies
Victor Christianto, Florentin Smarandache & Yunita Umniyati   79-92

Opinion

The Impossibility of Direct Detection of Big Bang Relic Neutrino
Victor Christianto & Florentin Smarandache   93-95

Commentary

Unidirectional Beams
B. G. Sidharth   96-96

Article

Derivation of the Muon *g*-2 Anomaly from Non-Equilibrium Dynamics

Ervin Goldfain[*]

Advanced Technology and Sensor Group, Welch Allyn Inc., Skaneateles Falls, NY 13153

Abstract

As it is known, the anomalous magnetic moment of muons is a longstanding puzzle of the Standard Model (SM). Commonly referred to as the muon "g-2 problem", the precise testing of this anomaly provides a sensitive probe for physics beyond SM. Here we show that the leading order contribution to the muon anomaly can be estimated from the onset of non-equilibrium dynamics near the Fermi scale. The derivation is straightforward and evades the postulated existence of new phenomena in the low to mid TeV range of high-energy physics.

Keywords: g-2 anomaly, Standard Model, minimal fractal manifold, Fractional Field Theory, non-equilibrium dynamics.

1. Introduction

In Quantum Mechanics, a charged particle of mass m carries a magnetic dipole $\boldsymbol{\mu}$ linked with its spin vector $\boldsymbol{s}$ through the following relationship

$$\boldsymbol{\mu} = g\left(\frac{\pm e}{2m}\right)\boldsymbol{s} \tag{1}$$

in which g is the gyromagnetic ratio and e the electric charge. Whereas in Quantum Mechanics g holds a well-defined value ($g = 2$), the cumulative contribution of quantum fluctuations in the SM yields a minute but non-vanishing correction to the gyromagnetic ratio. This correction is parameterized by the so-called *anomalous magnetic moment,* which is defined as

$$a = \frac{g-2}{2} \tag{2}$$

Another useful parameterization of (2) is the deviation between its experimental and theoretical values, i.e.,

---

[*]Correspondence: Ervin Goldfain, Ph.D., Photonics CoE, Welch Allyn Inc., Skaneateles Falls, NY 13153, USA
E-mail: ervingoldfain@gmail.com

$$\Delta a = \left| a^{exp} - a^{th} \right| \tag{3}$$

When applied to the family of SM charged leptons (the electron e , muon μ and the tau-lepton τ), (3) can be measured with various degrees of accuracy and evaluated against all perturbative corrections demanded by the theory. The overall contribution of the SM may be split into three components, namely,

$$a_l^{SM} = a_l^{QED} + a_l^{EW} + a_l^{Had} \ , \quad l = \left\{ e, \mu, \tau \right\} \tag{4}$$

corresponding to the electrodynamics (QED), electroweak (EW) and the hadronic sectors, respectively. The leptonic deviation (3) assumes the form

$$\Delta a_l = \left| a_l^{exp} - a_l^{SM} \right| \tag{5}$$

Conventional wisdom holds that (5) can be tracked down to either an experimental error, an incorrect computation of hadronic loop-diagrams or a hint for "*new physics*" phenomena (NP), alleged to surface beyond the energy range of the SM. While the electron anomalous moment Δa_e is relatively insensitive to the EW and strong sectors of the SM, the muon anomalous moment Δa_μ contains contributions from all SM sectors. As a result, precise tests of Δa_μ are believed to provide an excellent opportunity to reveal or constrain the signature of NP, including, for example, supersymmetry, leptoquarks, extended gauge groups, seesaw models, extra dimensions, extended Higgs sectors and other field theories beyond SM (referred below to BSM scenarios) [1-3].
The most recent experimental values on the magnitude of Δa_e and Δa_μ are [1, 7]

$$\Delta a_e^{exp} = (-0.88) \times 10^{-12} \tag{6}$$

$$\Delta a_\mu^{exp} = 2.87(\pm 0.80) \times 10^{-9} \tag{7}$$

The study of lepton magnetic moments represents an active research subject in particle physics. For additional information and in-depth details, the reader is directed to the large database of articles and books devoted to this topic.

2. Non-equilibrium dynamics as source of the g-2 anomaly

The concept of *helicity* in Quantum Field Theory (QFT) refers to the projection of spin onto the direction of the momentum vector. A distinctive feature of the anomalous magnetic moment of leptons is that it can flip helicity in interactions mediated by gauge bosons [2]. These flip transitions are only allowed for massive particles with a probability amplitude proportional to the mass of the particle. Since the probability of such helicity flips scales with the square of the probability amplitude, the lepton anomalous magnetic moment Δa_l must scale with the square of the lepton mass. In fact, it can be shown that, in the presence of BSM physics, the leading one-loop contribution to the muon anomalous magnetic Δa_μ is on the order of [1]

$$\Delta a_\mu \sim \frac{g_{NP}^2}{16\pi^2}\frac{m_\mu^2}{M_{NP}^2} \tag{8}$$

where g_{NP} and M_{NP} denote the coupling and mass associated with the onset of the NP sector, in particular, *heavy BSM particles* alleged to show up beyond the energy range of the Large Hadron Collider. Many avenues centered around (8) have been explored from various vantage points. For instance, [1] examines the possibility of one or two new fields with different spins and gauge group representations, under the hypothesis of weakly coupling $|g_{NP}| \leq \sqrt{4\pi}$ and weak-scale masses $M_{NP} \geq 100\,\text{GeV}$.

In what follows we continue to use (8) as a baseline but interpret it in a new light. Specifically, our analysis proceeds with the following assumptions:

1) Rather than postulating that (8) follows from the NP sector of particle physics, we conjecture that (8) is rooted in the onset of *non-equilibrium dynamics near* the Fermi scale given by $M_{NP} \to M_{EW} \approx O(246\,\text{GeV})$ [4-6].

2) we consider the dominant muon decay channel known as the *Michel decay* [2],

$$\mu^- \to \nu_\mu e^- \overline{\nu}_e \tag{9}$$

which is a pure leptonic process mediated by the weak boson W^-. As such, (9) is a reliable process since it has high-statistics, it enables precise measurements of the Fermi constant(G_F) and it accounts for nearly 100% of the branching ratio.

Because we are limiting the discussion to the Michel decay, it makes sense to approximate the scale where non-equilibrium sets in as $O(M_{EW}) \approx M_W$. This is to say that the mass of the mediating W^- boson marks the representative non-equilibrium scale for (9).
The muon coupling in (9) can be extracted from the SM relationship

$$\frac{G_F}{\sqrt{2}} = \frac{g_\mu^2}{8M_W^2} \tag{10}$$

where M_W is the mass of the W boson. From (8) and (10) we derive

$$\Delta a_\mu \sim \frac{G_F m_\mu^2}{2\sqrt{2}\,\pi^2} \tag{11}$$

which yields

$$\boxed{\Delta a_\mu \sim 4.667 \times 10^{-9}} \tag{12}$$

in promising agreement with the experimental data (7).

A natural follow-up question is the following: Would the same analysis apply to the anomalous magnetic moments of electrons and tau-leptons? Assuming that this line of thinking is correct and that the muon coupling g_μ stays the same for electrons and tau-leptons, we use (8) to estimate the non-equilibrium mass scales that match existing data on Δa_e and Δa_μ [1-3, 7-8]. Numerical evaluation leads to

$$\Delta a_e \to \lambda M_W, \ \lambda = \frac{1}{3} \tag{13}$$

$$\Delta a_\tau \to \lambda M_W \approx M_{EW}, \ \lambda = 3 \tag{14}$$

It is also instructive to ponder on the possible corrections to the muon coupling due to the "would-be" contribution of lepton flavor violation (LFV) [2]. This analysis goes beyond the scope of the paper and may be considered elsewhere.

Our findings are summarized in the table below.

Table 1: Comparison between experimental and predicted values of Δa_l

	Experiment	**Prediction**
Δa_μ	$2.87(\pm 0.80) \times 10^{-9}$	4.667×10^{-9}
Δa_e	$(-0.88) \times 10^{-12}$	$(-0.605) \times 10^{-12}$ for $O(M_{EW}) = (\frac{1}{3}) M_W$
Δa_τ	$-0.0508 < \Delta a_\tau^{exp} < 0.011823$	$\sim 10^{-7}$ for $O(M_{EW}) = 3 M_W$

Received March 6, 2020; Accepted March 30, 2020

References

[1] https://arxiv.org/pdf/1402.7065.pdf

[2] https://arxiv.org/pdf/1610.06587.pdf

[3] https://arxiv.org/pdf/2002.04822.pdf

[4] Available at the following sites:
http://www.aracneeditrice.it/aracneweb/index.php/pubblicazione.html?item=9788854889972
https://www.researchgate.net/publication/278849474_Introduction_to_Fractional_Field_Theory_consolid
   ated_version

[5] http://www.ejtp.info/articles/ejtpv7i24p219.pdf

[6] Available at the following site:
https://www.researchgate.net/publication/266578650_REFLECTIONS_ON_THE_FUTURE_OF_PARTI
   CLE_THEORY

[7] https://science.sciencemag.org/content/360/6385/191

[8] https://iopscience.iop.org/article/10.1088/1742-6596/912/1/012001/pdf

Article

Solving the Flatness & Horizon Problems
via Self-Organized Criticality

Ervin Goldfain[*]

Advanced Technology and Sensor Group, Welch Allyn Inc., Skaneateles Falls, NY 13153

Abstract

Self-organized criticality (SOC) is a universal mechanism for self-sustained critical behavior in large-scale systems evolving outside equilibrium. The trademark signature of SOC is two-fold: a) it occurs in complex ensembles of multiple interacting components and b) it is characterized by power-law distribution of "avalanche" sizes. This brief report suggests that both flatness and horizon problems of cosmology may be explained away through the universal features of SOC. The explanation stems from the so-called *finite scaling ansatz* (FSS) of SOC, which is a generic paradigm for the emergence of complexity in Nature. Our approach is straightforward and evades traditional solutions involving fine-tuning, particle horizons or inflation.

Keywords: flatness problem, horizon problem, inflation, fine tuning, self-organized criticality, finite scaling ansatz, minimal fractal manifold.

1. From SOC to the minimal fractal geometry of spacetime

Consider a large-scale system of size L undergoing a second-order phase transition. The transition is driven by the control parameter λ as it approaches the critical value λ_c. Near the critical point and for systems of infinite extent ($L \to \infty$), the correlation length ξ diverges as [10-12]

$$\xi \sim (\lambda - \lambda_c)^{-\nu} \; ; \; L \to \infty, \; \lambda \to \lambda_c \tag{1}$$

In the transition region, a relevant variable of the system is also a diverging quantity which scales as

$$A_\infty(\lambda) \sim \left| \lambda - \lambda_c \right|^{-\varsigma} ; \; L \to \infty, \; \lambda \to \lambda_c \tag{2}$$

where ς is a critical exponent. In what follows, we introduce the notation

<hr>

[*]Correspondence: Ervin Goldfain, Ph.D., Photonics CoE, Welch Allyn Inc., Skaneateles Falls, NY 13153, USA
E-mail: ervingoldfain@gmail.com

$$\tau_s = -\left(\frac{\varsigma}{\nu}\right) \tag{3}$$

There are two distinct cases associated with the power-law (2). If the size of the system greatly exceeds the correlation length, $L \gg \xi$, by (1) and (2) we write

$$A_L(\lambda) \sim \xi^{-\tau_s} \; ; \; (L \gg \xi, \; \lambda \to \lambda_c) \tag{4}$$

In the opposite case, $L \ll \xi$, the system size takes over the scaling behavior and (2) turns into

$$A_L(\lambda) \sim L^{-\tau_s} \; ; \; (L \ll \xi, \; \lambda \to \lambda_c) \tag{5}$$

Taken together, (4) and (5) define the *finite-size scaling* (FSS) ansatz [1, 10-11]

$$A_L(\lambda) = \xi^{-\tau_s} \Phi\left(\frac{L}{\xi}\right) \; ; \; (L \to \infty, \; \lambda \to \lambda_c) \tag{6}$$

where the scaling function controls the finite-size effects of critical behavior and is defined as

$$\Phi(x) = \begin{cases} const; \; |x| \gg 1 \\ x^{-\tau_s} \; ; \; x \to 0 \end{cases} \tag{7}$$

To transition from the framework of critical phenomena to SOC, one simply identifies the correlation length with the concept of *avalanche-size*, i.e.,

$$s = \xi \; ; \quad s_{cr} = L \tag{8}$$

The probability distribution defining the FSS ansatz in SOC is a natural extrapolation of (6) and takes the form of a probability distribution [11]

$$\boxed{P(s,L) \sim s^{-\tau_s} \Phi\left(\frac{s}{s_c}\right) \text{ for } s \gg 1, \; L \gg 1} \tag{9a}$$

$$\boxed{s_c(L) \sim L^{D_0} \text{ for } L \gg 1} \tag{9b}$$

in which τ_s and D_0 are called the *avalanche-size exponent* and the *avalanche dimension*, respectively. Quite generally, (9) shows that, for a system of finite size and large size avalanches, the avalanche-size probability behaves as a fractal function times a generic scaling function. To enable all moments of (9) to exist, the scaling function must decay sufficiently fast. One obtains the following representation of the scaling function upon power expanding it around zero,

$$\Phi(x) \sim \begin{cases} \Phi(0) + \Phi'(0)x + \dfrac{1}{2}\Phi''(0)x^2 + ..., & x \ll 1 \\ \rightarrow 0, & x \gg 1 \end{cases} \qquad (10)$$

The avalanche-size probability must be normalized to unity and its average be diverging along with $L \rightarrow \infty$, which leads to the following constraints

$$\sum_{s=1}^{\infty} P(s;L) = 1 \qquad \text{for } L < \infty, \qquad (11)$$

$$\langle s \rangle = \sum_{s=1}^{\infty} sP(s;L) \rightarrow \infty \ \text{ for } L \rightarrow \infty \qquad (12)$$

Under the assumption that $\Phi(0) \neq 0$, the behavior of (9) for an infinite system size may be approximated as

$$\lim_{L \rightarrow \infty} P(s;L) \sim s^{-\tau_s}\Phi(0) \qquad (13)$$

Furthermore, to comply with (11) and (12), the avalanche-size exponent must fall in the range

$$1 < \tau_s \leq 2 \qquad (14)$$

We have extensively discussed in [6 - 9] the physical significance of the *minimal fractal manifold* (MFM), a spacetime continuum characterized by arbitrarily small and scale-dependent deviations from four dimensions ($\varepsilon = 4 - D \ll 1$). The MFM reflects an evolving setting that starts far-from-equilibrium and asymptotically reaches the equilibrium conditions mandated by field theory in the limit of four-dimensional spacetime ($\varepsilon = 0$). There are well-motivated reasons to believe that dimensional fluctuations driven by ε are asymptotically compatible with the internal structure and dynamics of the Standard Model of particle physics [6-9].
Based on these premises, we advance below the hypothesis that the dimensional deviation ε and the avalanche-size s are interchangeable concepts via

$$\varepsilon = 4 - D = s^{-1} \ll 1 \qquad (15)$$

Furthermore, since ε flows with the energy scale, it likely reaches its uppermost observable value close to the formation of the cosmic microwave background (CMB) [13]. Thus, the maximal dimensional deviation is set to

$$\varepsilon_{cr} = \varepsilon_{\max} \approx 10^{-5} \ ; \ \varepsilon \ll \varepsilon_{cr} \qquad (16)$$

which turns (9) into

$$P(\varepsilon, \varepsilon_{cr}) \sim \varepsilon^{\tau_s} \Phi(\varepsilon_{cr}/\varepsilon) \ , \ \varepsilon \ll 1 \tag{17a}$$

$$\varepsilon_{cr}(a) \sim a^{D_0} \ , \ a \gg 1 \tag{17b}$$

where $a = a(t)$ is the scale factor describing the Universe expansion.

Next paragraph deploys these ideas towards solving the flatness and horizon problems, two major topics of contemporary cosmology [2-5].

2. The asymptotic approach to flatness and homogeneity

The considerations outlined so far suggest that the large-scale dynamics of the Universe may be naturally interpreted as a *global SOC process*. In light of this viewpoint, an evolving cosmological parameter – be it the deviation from spacetime flatness or the homogeneity across causally disconnected patches of the Universe – asymptotically approaches a quasi-stationary value representing a *non-equilibrium steady state (NESS)*. Moreover, it is conceivable that, while dimensional fluctuations induced by $\varepsilon = 4 - D \ll 1$ provide the *driving* mechanism of cosmological SOC, Universe expansion acts as a *dissipation* reservoir. One may reasonably infer from (9) and (15) - (17) that the observed value of such a parameter $A_{a(t)}(\varepsilon)$ flows with ε as in

$$A_{a(t)}(\varepsilon) = A_{a_0}(\varepsilon)[1 - \varepsilon^{\tau_s} \Phi(\varepsilon/\varepsilon_{cr})] \ , \ \varepsilon \ll 1, \ a(t) \gg 1 \tag{18}$$

in which a_0 denotes the scale factor associated with the NESS of Universe expansion. Scaling (18) can be cast in the equivalent form

$$\frac{\Delta A_{a_0}(\varepsilon)}{A_{a_0}(\varepsilon)} = \frac{A_{a_0}(\varepsilon) - A_{a(t)}(\varepsilon)}{A_{a_0}(\varepsilon)} = \varepsilon^{\tau_s} \Phi(\varepsilon/\varepsilon_{cr}) \tag{19}$$

or

$$\frac{\Delta A_{\infty}(s)}{A_{\infty}(s)} = \varepsilon^{\tau_s} \overline{\Phi}(\varepsilon_{cr}/\varepsilon) \tag{20}$$

where

$$\overline{\Phi}(\varepsilon_{cr}/\varepsilon) = \Phi(\varepsilon/\varepsilon_{cr}) = \Phi[(\varepsilon_{cr}/\varepsilon)^{-1}] \tag{21}$$

Power expanding (21) by analogy with (10) yields

ISSN: 2153-8301　　　　　　　　　　Prespacetime Journal　　　　　　　　www.prespacetime.com
Published by QuantumDream, Inc.

$$\overline{\Phi}(x) \sim \begin{cases} \overline{\Phi}(0) + \overline{\Phi}'(0)\,x + \dfrac{1}{2}\overline{\Phi}''(0)\,x^2 + ..., & x \ll 1 \\ \to 0, & x \gg 1 \end{cases}$$

which leads to

$$\boxed{\Delta A_{a_0}(\varepsilon) = 0 \; ; \; \overline{\Phi}\left(\varepsilon_{cr}\big/\varepsilon\right) \to 0, \; \varepsilon \ll \varepsilon_{cr}, \; \varepsilon \to 0} \qquad (22)$$

It is apparent from (22) that the end-state of the asymptotic approach to flatness and homogeneity matches the classical limit of four-dimensional spacetime ($D = 4$).

Received April 3, 2020; Accepted May 10, 2020

References

1. Available at the following site:
   https://pdfs.semanticscholar.org/9c4b/cc2faf5729cb74e4428f5c109d18c41b10be.pdf?_ga=2.10769587
   .687731597.1585061445-274809731.1569191893

2. Available at the following site:
http://jamesowenweatherall.com/wp-content/uploads/2014/10/What-Is-the-Horizon-Problem.pdf

3. https://arxiv.org/pdf/1803.05148.pdf

4. http://icc.ub.edu/~liciaverde/TALKS/cosmo5.pdf

5. https://arxiv.org/pdf/1911.02828.pdf

6. Available at the following sites:
http://www.aracneeditrice.it/aracneweb/index.php/pubblicazione.html?item=9788854889972
https://www.researchgate.net/publication/278849474_Introduction_to_Fractional_Field_Theory_consolid
   ated_version

7. http://www.ejtp.info/articles/ejtpv7i24p219.pdf

8. Available at the following site:
https://www.researchgate.net/publication/266578650_REFLECTIONS_ON_THE_FUTURE_OF_PARTI
   CLE_THEORY

9. References [2 – 12] listed in
https://www.prespacetime.com/index.php/pst/article/view/1634/1564

10. https://pyfssa.readthedocs.io/en/stable/fss-theory.html

11. Christensen, K. and Moloney, N. R., "Complexity and Criticality", Imperial College Press 2005.

12. Amit, D. J. and Martin-Mayor, V., "Field Theory, the Renormalization Group and Critical
    Phenomena", World Scientific 2005.

13. https://arxiv.org/pdf/0806.2675.pdf

Article

Quantum Field Theory as Manifestation of Self-Organized Criticality

Ervin Goldfain[*]

Advanced Technology and Sensor Group, Welch Allyn Inc., Skaneateles Falls, NY 13153

Abstract

Self-organized criticality (SOC) reflects the ability of many complex dynamical systems to self-sustain critical behavior outside equilibrium. Here we provide analytic evidence that quantum field propagators and the probability distribution of SOC share a common foundation. In particular, we find that the formal structure of quantum propagators replicates the *finite scaling ansatz* (FSS) of SOC, which is a generic paradigm for the emergence of complexity in Nature.

Keywords: complexity, self-organized criticality, non-equilibrium dynamics, finite size scaling ansatz, quantum field propagator.

1. SOC and the FSS ansatz

Consider a large-scale system of size L undergoing a second-order phase transition. The transition is driven by the control parameter λ as it approaches the critical value λ_c. Near the critical point and for systems of infinite extent ($L \to \infty$), the correlation length ξ diverges as [1-3]

$$\xi \sim (\lambda - \lambda_c)^{-\nu} \; ; \; L \to \infty, \; \lambda \to \lambda_c \tag{1}$$

In the transition region, a relevant variable of the system is also a diverging quantity which scales as

$$A_\infty(\lambda) \sim |\lambda - \lambda_c|^{-\zeta} \; ; \; L \to \infty, \; \lambda \to \lambda_c \tag{2}$$

where ζ is a critical exponent. In what follows, we introduce the notation

[*]Correspondence: Ervin Goldfain, Ph.D., Photonics CoE, Welch Allyn Inc., Skaneateles Falls, NY 13153, USA
E-mail: ervingoldfain@gmail.com

$$\tau_s = -\left(\frac{\varsigma}{\nu}\right) \tag{3}$$

There are two distinct cases associated with the power-law (2). If the size of the system greatly exceeds the correlation length, $L >> \xi$, by (1) and (2) we write

$$A_L(\lambda) \sim \xi^{-\tau_s} \; ; \; (L >> \xi, \; \lambda \to \lambda_c) \tag{4}$$

In the opposite case, $L << \xi$, the system size takes over the scaling behavior and (2) turns into

$$A_L(\lambda) \sim L^{-\tau_s} \; ; \; (L << \xi, \; \lambda \to \lambda_c) \tag{5}$$

Taken together, (4) and (5) define the *finite-size scaling* (FSS) ansatz [1-2]

$$A_L(\lambda) = \xi^{-\tau_s} \Phi\left(\frac{L}{\xi}\right) \; ; \; (L \to \infty, \; \lambda \to \lambda_c) \tag{6}$$

where the cutoff function controls the finite-size effects of critical behavior and is defined as

$$\Phi(x) = \begin{cases} const; \; |x| >> 1 \\ x^{-\tau_s} \; ; \; x \to 0 \end{cases} \tag{7}$$

To transition from the framework of critical phenomena to SOC, one simply identifies the correlation length with the concept of *avalanche-size*, i.e.,

$$s = \xi \; ; \quad s_c = L \tag{8}$$

The probability distribution defining the FSS ansatz in SOC is a natural extrapolation of (6) and takes the form [2]

$$\boxed{P(s,L) \sim s^{-\tau_s} \Phi\left(\frac{s}{s_c}\right) \text{ for } s >> 1, \; L >> 1} \tag{9a}$$

$$\boxed{s_c(L) \sim L^{D_0} \text{ for } L >> 1} \tag{9b}$$

in which τ_s and D_0 are called the *avalanche-size exponent* and the *avalanche dimension*, respectively. Quite generally, (9) shows that, for a system of finite extent and large size avalanches, the avalanche-size probability behaves as a power-law weighted by a cutoff function. To enable all moments of (9) to exist, the cutoff function must decay sufficiently fast. One obtains the following representation of the cutoff function upon power expanding it around zero,

$$\Phi(x) \sim \begin{cases} \Phi(0) + \Phi'(0)x + \dfrac{1}{2}\Phi''(0)x^2 + ..., & x \ll 1 \\[2mm] \to 0, & x \gg 1 \end{cases} \tag{10}$$

The avalanche-size probability must be normalized to unity and its average be diverging along with $L \to \infty$, which leads to the following constraints

$$\sum_{s=1}^{\infty} P(s;L) = 1 \qquad \text{for } L < \infty, \tag{11}$$

$$\langle s \rangle = \sum_{s=1}^{\infty} sP(s;L) \to \infty \ \text{ for } L \to \infty \tag{12}$$

Under the assumption that $\Phi(0) \neq 0$, the behavior of (9) for an infinite system size may be approximated as

$$\lim_{L \to \infty} P(s;L) \sim s^{-\tau_s} \Phi(0) \tag{13}$$

Furthermore, to comply with (11) and (12), the avalanche-size exponent must fall in the range

$$1 < \tau_s \leq 2 \tag{14}$$

We proceed next to the analysis of four examples linking the FSS ansatz to the formalism of quantum propagators. Before doing so, it is worth emphasizing that SOC and the FSS ansatz are deeply linked to *multifractals* and the *path integral* approach to quantum theory [14-16].

2. *Case #1*: free scalar relativistic particle

Let us begin by recalling the Klein-Gordon (KG) description of a relativistic particle of mass m in 3+1 spacetime dimensions. It is known that the perturbative treatment of the KG theory may be built by analogy with the Gaussian random walk (RW) model [4]. In the Euclidean version of this theory, ordinary time is analytically continued to $t = i\tau$ and the momentum-space propagator takes the form

$$G(r,p) = \frac{1}{\mathbf{p}^2 + p_0^2 + m^2} = \int_0^{\infty} dr \exp(-rm^2)\exp[-r(\mathbf{p}^2 + p_0^2)] \tag{15}$$

ISSN: 2153-8301   Prespacetime Journal   www.prespacetime.com
                            Published by QuantumDream, Inc.

where $p^\mu = (\mathbf{p}, p_0)$, with the energy analytically continued to $p_0 = -iE$. Fourier transforming the second integrand in (15) gives the probability distribution of Gaussian random walks of step length r,

$$p(\Delta, r) = (4\pi r)^{-2} \exp[-\Delta(r)] \tag{16}$$

in which

$$\Delta(r) = \frac{\rho^2}{4r} = \frac{|\mathbf{x}|^2 + \tau^2}{4r} \tag{17}$$

Since r has units of inverse mass squared, $r = [M]^{-2} = [L]^2$, it is convenient to normalize (16) through the substitution

$$r^0 = r m^2 \tag{18}$$

which turns (16) into

$$p^0(\Delta, r^0) = \frac{p(\Delta, r)}{m^4} = (4\pi r^0)^{-2} \exp(-\frac{\rho^2 m^2}{4 r m^2}) \tag{19}$$

or,

$$p^0(\Delta, r^0) = (4\pi r^0)^{-2} \exp(-\frac{\rho^2 m^2}{4 r^0}) = (4\pi r^0)^{-2} \exp[-\Delta(r)] \tag{20}$$

Comparing (20) and (17) with the FSS ansatz (9) gives

$$\boxed{\tau_s = 2, \ |D_0| = 2} \tag{21}$$

under the following assumptions:

a) the RW step length r^0 is the analogue of the avalanche size s introduced in (9),

b) the scaling dimension of ρ^2 in (17) is $\left[\rho^2\right] = [L]^2$, which implies $|D_0| = 2$.

3. *Case #2*: harmonic oscillator

Consider now the general case of an *anharmonic oscillator* in 1+1 dimensions with Euclidean Lagrangian

$$L_E = \frac{m\dot{x}^2}{2} + \frac{m\omega^2 x^2}{2} + \Delta V(x) = \overline{L_E} + \Delta V(x) \tag{22}$$

where $\overline{L_E}$ denotes the unperturbed Lagrangian and $\Delta V(x)$ a small perturbation to the dynamics of $\overline{L_E}$. Assuming natural units throughout ($\hbar = k = 1$), the path integral for the propagator is given by [5]

$$K_E[\beta, J] = \exp[-\int d\tau \Delta V(\frac{\delta}{\delta J(\tau)})]\overline{K_E}[\beta, J] \tag{23}$$

where $\overline{K_E}(\beta, J)$ represents the path integral of the harmonic oscillator, whose Euclidean Lagrangian $\overline{L_E}$ assumes the form

$$\overline{L_E} = \frac{m\dot{x}^2}{2} + \frac{m\omega^2 x^2}{2} \tag{24}$$

In (23), $t = -i\beta$ is the expression of time as function of the reduced temperature parameter of statistical mechanics ($\beta = T^{-1}$) and $\tau = it$ is the Euclidean time. Furthermore, $\overline{K_E}(\beta, J)$ is the Gaussian integral representing the Euclidean harmonic oscillator driven by the forcing function J. It can be shown that $\overline{K_E}(\beta, J)$ amounts to [5]

$$\overline{K_E}(\beta, J) = [\frac{m\omega}{2\pi \sinh(\omega\beta)}]^{\frac{1}{2}} \exp(J \cdot G_D \cdot J) \tag{25}$$

in which $G_D(\tau_1, \tau_2)$ stands for the Green function of the harmonic oscillator, computed between the initial (τ_1) and final (τ_2) times. For $\beta \to \infty$, ($T = 0$) the Green function reduces to

$$\lim_{\beta \to \infty} G_D = \frac{1}{2m\omega} \exp(-\omega|\tau_1 - \tau_2|) \tag{26}$$

if one of these two conditions are satisfied

$$|\tau_1 - \tau_2| << \beta \tag{27a}$$

$$\left| \tau_{1,2} \pm \frac{\beta}{2} \right| << \beta \tag{27b}$$

The transition probability between the initial and final times is the square of (26), that is,

$$P_{D(1,2)} = G_D^2 = (\frac{1}{2m\omega})^2 \exp(-\frac{2\omega}{|\tau_1 - \tau_2|^{-1}}) \tag{28}$$

To express (28) in a normalized form, we introduce the dimensionless parameter

$$r^0 = m\omega\beta^2 \tag{29}$$

which turns the propagator (26) and transition probability (28) into, respectively,

$$G_D^0 = \frac{G_D}{\beta^2} = \frac{1}{2r^0}\exp(-\omega|\tau_1 - \tau_2|) = \frac{1}{2r^0}\exp(-\frac{r^0}{m\beta^2}|\beta_1 - \beta_2|) \tag{30}$$

$$P_{D(1,2)}^2 = (\frac{1}{2r^0})^2 \exp(-\frac{2r^0}{m\beta^2}|\beta_1 - \beta_2|) \tag{31}$$

Since the scaling dimension of the Euclidean time and of the inverse temperature parameter is $[\tau] = [\beta] = [L]^1 = [M]^{-1}$, by (9) and (31) we obtain

$$\boxed{\tau_s = 2, \ |D_0| = 1} \tag{32}$$

4. *Case #3*: oscillator with quartic interaction

Let us look next at the case of a massless scalar field with quartic self-interaction, coupled to an external current J [6]. The generating functional and propagator are respectively given by

$$Z[J] = \int [d\varphi] \exp\{i \int d^4x [\frac{1}{2}(\partial\varphi)^2 - \frac{\lambda^4}{4}\varphi^4 + J\varphi]\} \tag{33}$$

$$G(t_1 - t_2) = \theta(t_1 - t_2)\mu(\sqrt[4]{2}\,\lambda^{-1})f(\lambda, \mu t) \tag{34}$$

$$f(\lambda, \mu t) = \mathrm{sn}(\frac{\lambda}{\sqrt[4]{2}}\mu t) \tag{35}$$

ISSN: 2153-8301   Prespacetime Journal   www.prespacetime.com
Published by QuantumDream, Inc.

where $\theta(t)$ is the time-ordering operator, μ an arbitrary constant with the dimension of a mass (that is, $[M]^1$) and $sn(...)$ stands for the Jacobi elliptic function. The dimensionless version of (34) reads

$$G^0(t_1 - t_2) = \frac{G(t_1 - t_2)}{\mu} = \theta(t_1 - t_2)(\sqrt[4]{2}\,\lambda^{-1})f(\lambda, \mu t) \tag{36}$$

Note that, because λ is a real number, it can be presented as power of another arbitrary real number, which indicates that the value of the avalanche-size exponent is *undetermined*. In line with previous arguments, since time t and energy scale μ entering (36) have scaling dimension ± 1, the oscillator with quartic interaction is characterized by

$$\boxed{\tau_s = undetermined, \; |D_0| = 1} \tag{37}$$

5. *Case #4*: Dirac theory in 1+1 dimensions

In this last example we examine the fundamental solution of the Dirac equation in 1+1 dimensions, which can be written as ($\hbar = c = 1$) [7]

$$\begin{pmatrix} -m & i\partial/\partial t - i\partial/\partial x \\ i\partial/\partial t + i\partial/\partial x & -m \end{pmatrix} \begin{pmatrix} \psi_1 \\ \psi_2 \end{pmatrix} = 0 \tag{38}$$

The corresponding propagator evaluated inside the future light cone $t > |x|$ assumes the form

$$K(x,t) = \frac{m}{2} \begin{pmatrix} -\dfrac{t+x}{\sqrt{t^2 - x^2}} J_1(m\sqrt{t^2 - x^2}) & iJ_0(m\sqrt{t^2 - x^2}) \\ iJ_0(m\sqrt{t^2 - x^2}) & \dfrac{-t+x}{\sqrt{t^2 - x^2}} J_1(m\sqrt{t^2 - x^2}) \end{pmatrix} \tag{39}$$

where each matrix entry includes the Bessel functions $J_0(...)$ and $J_1(...)$. Analysis shows that the factor multiplying the Bessel functions in (39) has dimension $|L|^{-1} = [M]^1$, hence normalizing the propagator using a large scale $M_0 \gg m$ yields

$$K^0(x,t) = \frac{K(x,t)}{M_0} \tag{40}$$

Squaring (40) and repeating the previous line of arguments leads to

$$\boxed{\tau_s = 2, \; |D_0| = 1}$$
(41)

The next section attempts to bridge the gap between SOC and the minimal fractality of spacetime geometry near or above the Fermi scale.

6. From SOC to the minimal fractality of spacetime

We have extensively discussed in [8 - 11] the physical significance of the *minimal fractal manifold* (MFM), a spacetime continuum characterized by arbitrarily small and scale-dependent deviations from four dimensions ($\varepsilon = 4 - D \ll 1$). The MFM reflects an evolving setting that starts far-from-equilibrium and gradually reaches the equilibrium conditions mandated by field theory in the limit of four-dimensional spacetime ($\varepsilon = 0$). There are well-motivated reasons to believe that dimensional fluctuations driven by ε are asymptotically compatible with the internal structure and dynamics of the Standard Model of particle physics [8-11].

Based on these premises, we introduce the hypothesis that the dimensional deviation ε and the avalanche-size s are interchangeable concepts via

$$\varepsilon = 4 - D = s^{-1} \ll 1$$
(42)

Furthermore, since ε flows with the energy scale, it likely reaches its uppermost observable value close to the formation of the cosmic microwave background (CMB) [12]. The maximal dimensional deviation is therefore set to

$$\varepsilon_c = \varepsilon_{\max} \approx 10^{-5} \; ; \; \varepsilon \ll \varepsilon_c$$
(43)

which turns (9) into

$$P(\varepsilon, \varepsilon_c) \sim \varepsilon^{\tau_s} \, \Phi\left(\varepsilon_c / \varepsilon\right) \, , \; \varepsilon \ll 1$$
(44a)

$$\varepsilon_c(\mu) \sim \mu^{D_0}, \; \mu \gg 1$$
(44b)

where μ is the dimensionless Renormalization Group scale. Using (44) as a baseline, a planned extension of this work will begin exploring the non-equilibrium regime of vacuum fluctuations, beyond the boundaries of perturbative Quantum Field Theory [13].

Received April 15, 2020; Accepted May 10, 2020

References

1. https://pyfssa.readthedocs.io/en/stable/fss-theory.html

2. Christensen, K. and Moloney, N. R., "Complexity and Criticality", Imperial College Press 2005.

3. Amit, D. J. and Martin-Mayor, V., "Field Theory, the Renormalization Group and Critical Phenomena", World Scientific 2005.

4. https://arxiv.org/pdf/1210.2630.pdf

5. http://www.weizmann.ac.il/particle/perez/Courses/QMII15/PInotes_Rattazzi.pdf (chapter 2).

6. https://arxiv.org/pdf/0807.4299.pdf

7. https://pdfs.semanticscholar.org/5379/148d876118d57e311c24b98eeffbc61cb191.pdf

8. Available at the following sites:
http://www.aracneeditrice.it/aracneweb/index.php/pubblicazione.html?item=9788854889972

https://www.researchgate.net/publication/278849474_Introduction_to_Fractional_Field_Theory_consolid ated_version

9. http://www.ejtp.info/articles/ejtpv7i24p219.pdf

10. Available at the following site:
https://www.researchgate.net/publication/266578650_REFLECTIONS_ON_THE_FUTURE_OF_PARTI CLE_THEORY

11. References [2 – 12] listed in
https://www.prespacetime.com/index.php/pst/article/view/1634/1564

12. https://arxiv.org/pdf/0806.2675.pdf

13. E. Goldfain, "Emergence of Lagrangian Field Theory from Self-Organized Criticality" (*under construction*).

14. Available at the following site:
https://pdfs.semanticscholar.org/9c4b/cc2faf5729cb74e4428f5c109d18c41b10be.pdf?_ga=2.10769587.68 7731597.1585061445-274809731.1569191893

15. https://arxiv.org/pdf/cond-mat/9805045.pdf

16. https://www.prespacetime.com/index.php/pst/article/view/1142/1143

ISSN: 2153-8301   Prespacetime Journal   www.prespacetime.com
Published by QuantumDream, Inc.

Article

Emergence of Lagrangian Field Theory from Self-Organized Criticality

Ervin Goldfain[*]

Advanced Technology and Sensor Group, Welch Allyn Inc., Skaneateles Falls, NY 13153

Abstract

Self-organized criticality (SOC) is a universal mechanism for self-sustained critical behavior in large-scale systems evolving outside equilibrium. Our report explores a tentative link between SOC and Lagrangian field theory, with the long-term goal of bridging the gap between complex dynamics and the non-perturbative behavior of quantum fields.

Keywords: Self-organized criticality, multifractals, Lagrangian field theory, non-equilibrium dynamics, complexity theory.

1. Key concepts of Self-Organized Criticality

The study of equilibrium critical phenomena reveals that, near a second-order phase transition, the scaling behavior of physical observables follows the so-called *finite-size scaling (FSS) ansatz*. By analogy, the probability distribution defining the FSS ansatz in SOC takes the form [1, 5-6, 9-10, 13]

$$P(s,L) \sim s^{-\tau_s} \Phi\left(\frac{s}{s_c} \right) \text{ for } s \gg 1,\ L \gg 1 \tag{1a}$$

$$s_c(L) \sim L^{D_0} \text{ for } L \gg 1 \tag{1b}$$

in which s_c is the *cutoff in the avalanche-size* and where τ_s and D_0 are called the *avalanche-size exponent* and the *avalanche dimension*, respectively. Quite generally, (1) shows that, for a system of finite extent and large avalanches, the avalanche-size probability behaves as a power-law weighted by a cutoff function. By analogy with equilibrium critical phenomena, the couple of exponents τ_s and D_0 determine the *universality class* of the SOC model.

---

[*]Correspondence: Ervin Goldfain, Ph.D., Photonics CoE, Welch Allyn Inc., Skaneateles Falls, NY 13153, USA
E-mail: ervingoldfain@gmail.com

To enable all moments of (1) to exist, the cutoff function must decay sufficiently fast. One obtains the following representation of the cutoff function upon power expanding it around zero [5],

$$\Phi(x) \sim \begin{cases} \Phi(0) + \Phi'(0)x + \dfrac{1}{2}\Phi''(0)x^2 + ..., & x << 1 \\[2mm] \rightarrow 0, & x >> 1 \end{cases} \tag{2}$$

The avalanche-size probability must be normalized to unity and its average be diverging along with $L \rightarrow \infty$, which leads to the following constraints

$$\sum_{s=1}^{\infty} P(s;L) = 1 \qquad \text{for } L < \infty, \tag{3}$$

$$\langle s \rangle = \sum_{s=1}^{\infty} s P(s;L) \rightarrow \infty \ \text{ for } L \rightarrow \infty \tag{4}$$

Under the assumption that $\Phi(0) \neq 0$, the behavior of (1) for an infinite system size may be approximated as

$$\lim_{L \rightarrow \infty} P(s;L) \sim s^{-\tau_s} \Phi(0) \tag{5}$$

Furthermore, to comply with (3) and (4), the avalanche-size exponent must fall in the range

$$1 < \tau_s \leq 2 \tag{6}$$

One may transition from the framework of equilibrium critical phenomena to SOC under the plausible assumption that the correlation length ξ displays the same behavior as the avalanche-size, i.e.,

$$s = \xi \ ; \quad s_c = L \tag{7}$$

2. From SOC to the minimal fractal manifold (MFM)

Refs. [11-12, 14, 19-22] have discussed at length the physical significance of the *minimal fractal manifold* (MFM), a spacetime continuum characterized by arbitrarily small and scale-dependent deviations from four dimensions ($\varepsilon = 4 - D << 1$). The MFM reflects an evolving setting that starts far-from-equilibrium and gradually reaches the equilibrium conditions mandated by field theory in the limit of four-dimensional spacetime ($\varepsilon = 0$). There are well-motivated reasons to believe that dimensional fluctuations driven by ε are asymptotically compatible with the internal structure and dynamics of the Standard Model of particle physics [11-12, 14, 19-22].

Based on these premises, as well as on (7), we introduce the hypothesis that the dimensional deviation ε and the avalanche-size $s = \xi$ are interchangeable concepts via

$$\varepsilon = 4 - D = s^{-1} << 1 \tag{8}$$

This hypothesis is consistent with the philosophy of *conformal field theory*, as the four-dimensional limit $\varepsilon = 0$, $(D = 4)$ naturally matches the asymptotic approach to the far-infrared regime of massless fields $(m = \xi^{-1} = s^{-1} \rightarrow 0)$.

Furthermore, since ε is conjectured to flow with the energy scale, it likely reaches its uppermost observable value close to the formation of the cosmic microwave background (CMB) [16]. The maximal dimensional deviation is therefore set to

$$\varepsilon_c = \varepsilon_{max} \approx 10^{-5} \; ; \; \varepsilon << \varepsilon_c \tag{9}$$

which turns (1) into

$$P(\varepsilon, \varepsilon_c) \sim \varepsilon^{\tau_s} \Phi(\varepsilon_c / \varepsilon) \; , \; \varepsilon << 1 \tag{10a}$$

$$\varepsilon_c(\mu) \sim \mu^{D_0} , \; \mu >> 1 \tag{10b}$$

where μ is the dimensionless Renormalization Group scale.

It can be shown that, in general, the FSS ansatz (1) or (10) can be characterized through a set of critical exponents α defined through [2, 9]

$$\varepsilon_c^{f(\alpha)} = \int_{\varepsilon_c^\alpha}^\infty P(\varepsilon, \varepsilon_c) d\varepsilon \tag{11}$$

or,

$$f(\alpha) = \frac{\log \int_{\varepsilon_c^\alpha}^\infty P(\varepsilon, \varepsilon_c) d\varepsilon}{\log \varepsilon_c} \tag{12}$$

The real numbers α represent the so-called *Lipschitz-Hölder* (LH) exponents and the multifractal spectrum $f(\alpha)$ quantifies their continuous distribution in the range $\alpha \in [\alpha_{min}, \alpha_{max}]$.

As the two next sections show, the multifractal spectrum plays a crucial role in bridging the gap between SOC and Lagrangian field theory.

ISSN: 2153-8301   Prespacetime Journal   www.prespacetime.com
Published by QuantumDream, Inc.

3. Key concepts of multifractal analysis

Since multifractals and SOC are closely related, we now take a brief detour to delve into the topic of multifractal analysis.

From a mathematical standpoint, multifractal analysis is a theory of *self-similar measures* [2]. A *measure* is defined as function that assigns a number to certain subsets of a given set: the number is said to represent the measure of the set. The basic properties of measures are extensions of the geometrical concepts of *length, area and volume*, so that - for example - the measure of the union of two disjoint sets is the sum of the measures of the two sets, and the measure of the empty set is zero. Roughly speaking, a self-similar measure is a measure whose geometrical attributes stay unchanged upon arbitrary scaling operations.

Following [2-3] in detail, let a set Σ supporting a measure be covered with a collection of boxes of size $\varepsilon = 4 - D \ll 1$. The number of boxes needed to cover the set is defined through the scaling

$$N(\varepsilon) \sim \varepsilon^{-D_H} \tag{13}$$

in which D_H is the Hausdorff dimension of the set, which is adequate for characterization of *mono-fractals*. In general, the quantitative description of *multifractal* measures requires replacing D_H with a continuous LH exponent α according to

$$\mu \sim \varepsilon^{\alpha} , \quad 0 < \alpha \in \left[\alpha_{\min}, \alpha_{\max}\right] < \infty \tag{14}$$

The number of boxes of size ε having the LH exponent α is given by

$$N_{\varepsilon}(\alpha) \sim \varepsilon^{-f(\alpha)} \tag{15}$$

where the distribution of LH exponents follows the multifractal spectrum $f(\alpha)$. The meaning of (15) is that there are infinitely many subsets of boxes having the LH exponent α in the limit $\varepsilon \to 0$.

By analogy with equilibrium statistical mechanics and Quantum Field Theory, multifractal analysis is based on a *partition function* defined as

$$Z_q(\varepsilon) = \sum_{i=1}^{N(\varepsilon)} \mu_i^q, \quad q \in \mathbf{R} \tag{16}$$

By (14), the measures assigned to boxes $i = 1, 2, ..., N(\varepsilon)$ is $\mu_i = \varepsilon^{\alpha_i}$. Assuming that the number of boxes for which $\alpha < \alpha_i < \alpha + d\alpha$ is $N_{\varepsilon}(\alpha) d\alpha$, the contribution of the subset of boxes with $\alpha_i \in [\alpha, \alpha + d\alpha]$ to the partition function is $N_{\varepsilon}(\alpha)(\varepsilon^{\alpha})^q d\alpha$ and thus

$$Z_q(\varepsilon) = \int N_\varepsilon(\alpha)(\varepsilon^\alpha)^q \, d\alpha \tag{17}$$

By (15) and (17), we obtain

$$Z_q(\varepsilon) = \int \varepsilon^{\,q\alpha - f(\alpha)} d\alpha \tag{18}$$

In the limit of four-dimensional spacetime $\varepsilon \to 0$ $(D = 4)$, the prevailing contribution to the integral (18) arises from those values of the exponent α that minimize the sum $q\alpha - f(\alpha)$ [2]. Such minimum exists if

$$\boxed{\frac{\partial}{\partial\alpha}\{q\alpha - f(\alpha)\} = 0} \tag{19}$$

which implies two conditions, namely

$$q = \frac{\partial}{\partial\alpha} f(\alpha), \quad \alpha = \alpha(q) \tag{20}$$

and

$$\frac{\partial^2}{\partial\alpha^2} f(\alpha) < 0, \quad \alpha = \alpha(q) \tag{21}$$

Introducing the definition

$$\tau(q) = q\alpha(q) - f(\alpha(q)) \tag{22}$$

it can be shown that

$$\alpha(q) = \frac{\partial}{\partial q}[\tau(q) + K_1] \tag{23}$$

in which K_1 represents a constant independent of q, that is, $\partial K_1 / \partial q = 0$. By (20), relation (22) can be presented as

$$\boxed{\alpha(q) = \frac{\tau(q)}{q} + f(\alpha(q))\frac{\partial[\alpha(q) + K_2]}{\partial f(\alpha(q))}} \tag{24}$$

where K_2 is a constant independent of the multifractal spectrum, that is, $\partial K_2 / \partial f = 0$. Equations (22) (or (24)) and (23) give a parametric representation of the spectrum $f(\alpha)$ in terms of q.

Taken together, (22) and (23) act as a *Legendre transform* from the variables q and τ to the variables α and f.

The Legendre transform is frequently used in various areas of theoretical physics including classical mechanics, statistical mechanics and thermodynamics, as well as Quantum Field Theory (QFT) [15]. Next section sets up the connection between (24), on the one hand, and the Legendre transform of Lagrangian field theory, on the other.

4. From Multifractals to Lagrangian dynamics

Starting from (24), it can be shown that classical statistical mechanics offers a straightforward analog of multifractal analysis. With reference to Tab. 1, the temperature (T), internal energy (U), entropy (S) and free energy (F) are respectively echoed in multifractal theory by $q^{-1}, \alpha, f(\alpha)$ and $\tau(q)/q$ [2, 9]. Considering this analogy, the thermodynamic equation

$$F(T,V) = U(S,V) - TS \tag{25}$$

turns into a replica of the Legendre transform (24), where V stands for volume and

$$dU = TdS - pdV \Rightarrow T = \left(\frac{\partial U}{\partial S}\right)_V \tag{26}$$

Multifractals	Statistical Mechanics
q	$1/T$
$\alpha(q)$	U
$\tau(q)/q$	F/T
$f(\alpha)$	S

Tab 1: Mapping Multifractals to Statistical Mechanics
(T = temperature, U = energy, F = free energy, S = entropy)

It is well known that the dynamics of generic classical fields is governed by the least-action principle [8, 17]

$$S = \int d^4x \, L(\varphi, \partial_\mu \varphi) \tag{27}$$

where S is the action whose minimization yields the Euler-Lagrange equations

$$\delta S = 0 \Rightarrow \partial_\mu \frac{\delta L}{\delta(\partial_\mu \varphi)} - \frac{\delta L}{\delta \varphi} = 0 \tag{28}$$

The Hamiltonian associated with (27) is defined by

$$H(x) = \pi(x)\partial_0\varphi(x) - L(x) \tag{29}$$

in which the conjugate momentum is

$$\pi(x) = \frac{\delta L}{\delta(\partial_0\varphi)} \tag{30}$$

One can interpolate between Lagrangian and Hamiltonian formulation of the theory using the kinetic (T) and potential (V) components according to

$$L = T - V, \ H = T + V \Rightarrow L = H - 2V \tag{31}$$

By (31), if the potential is independent of $\partial_0\varphi$, (30) can be written as

$$\pi(x) = \frac{\delta L}{\delta(\partial_0\varphi)} = \frac{\delta H}{\delta(\partial_0\varphi)}, \ \ \text{if } \frac{\delta V}{\delta(\partial_0\varphi)} = 0 \tag{32}$$

Inspection of the first entry in Tab. 1 hints to the analogy

$$q \Leftrightarrow |t_E| = \frac{1}{T} \tag{33}$$

where $t_E = -ix_0$ is interpreted as *Euclidean time*, identical with the inverse temperature of statistical mechanics expressed in natural units $(k = 1)$ [18]. Side by side evaluation of (29) to (32) with (22) to (24) leads to the following operational identification

$$\partial_0 \Leftrightarrow \partial_{t_E} \tag{34}$$

$$\alpha(q) \Leftrightarrow H \tag{35}$$

$$-\frac{\tau(q)}{q} \Leftrightarrow L \tag{36}$$

$$\partial_0\varphi \Leftrightarrow f(\alpha) \tag{37}$$

$$\varphi \Leftrightarrow \varphi(\alpha) = \int f(\alpha)d|t_E| \approx \langle f(\alpha)\rangle \cdot |t_E| \tag{38}$$

$$K_1 = K_2 \Leftrightarrow -2V \tag{39}$$

Therefore $\varphi(\alpha)$ and $f(\alpha)$ form a pair of conjugate variables defining the Lagrangian field analog of SOC. Relations (33)-(37) are consolidated in Tab. 2.

Multifractals	Lagrangian dynamics
q	$\lvert t_E \rvert$
$\alpha(q)$	H
$\tau(q)/q$	$-L$
$f(\alpha)$	$\partial_0 \varphi$

Tab 2: Mapping Multifractals to Lagrangian dynamics

(t_E = Euclidean time, H = Hamiltonian, L = Lagrangian)

A key observation is now in order. Integrating (36) over the Euclidean time (33) shows that

$$S = -\int \left(\frac{\tau(\lvert t_E \rvert)}{\lvert t_E \rvert} \right) d^3x \, d\lvert t_E \rvert \approx -\left\langle \frac{\tau(\lvert t_E \rvert)}{\lvert t_E \rvert} \right\rangle \cdot \Delta^3 \cdot \lvert t_E \rvert \tag{40}$$

in which Δ denotes the spatial domain of integration. Taken together, (28) and (40) hint that *multifractal geometry and SOC lie at the root of the least-action principle.*

We close by noting that the sole intent of our work is to shed light on the tantalizing connection between Lagrangian dynamics and SOC. Future research may focus on several aspects of field theory that were not covered here, such for example, the propagator structure of QFT and the FSS ansatz (1) [7]. The reader is also directed to [14] for a preliminary analysis of the relationship between the Legendre transform and the path integral formalism of perturbative QFT.

Appendix

A closer inspection of (15) reveals that an arbitrary change of $f(\alpha)$ is formally equivalent to a shift in $\varepsilon = 4 - D \ll 1$, in response to a change of the dimensionless RG scale μ (refer to critical behavior in continuous dimension). To derive the connection between the two scaling relationships, we require the following alternative definition of (15) to hold true

$$N_\varepsilon(\alpha) = \varepsilon(\mu_0)^{-f(\alpha)} = \varepsilon(\mu)^{-f(\alpha_0)} \tag{A1}$$

where μ_0 and α_0 denote two reference values. (A1) leads to

$$\varepsilon(\mu) = \varepsilon(\mu_0)^{\sigma(\alpha, \alpha_0)} \tag{A2}$$

in which

$$\sigma(\alpha, \alpha_0) = \frac{f(\alpha)}{f(\alpha_0)} \tag{A3}$$

Consider next the simple case of a free particle of mass m_0 moving at low speed with linear momentum p in one space dimension. The dispersion relationship linking the linear momentum with energy is

$$m_0 = \frac{p^2}{2E} \tag{A4}$$

Passing to the Euclidean representation $E_E = -iE$ in (A4) yields the Euclidean mass $m_{0,E}$ in the form

$$m_{0,E} = -\frac{ip^2}{2E_E} \tag{A5}$$

When translated in terms of (35) and (37), the above equation suggests that the counterpart of Euclidean mass in the SOC framework is given by

$$\left| m_{0,E} \right| = \frac{f^2(\alpha)}{2\alpha} \tag{A6}$$

(A6) represents a linear function in the LH exponent α, if the spectrum $f(\alpha)$ is assumed to be quadratic. (A2) and (A6) hint that the dynamical property of "mass" emerges from the *continuous dimensionality of spacetime* $\varepsilon = 4 - D \ll 1$, an argument made throughout our previous publications (see for example [11-12]).

Received April 16, 2020; Accepted May 16, 2020

References

1. https://arxiv.org/pdf/cond-mat/9805045.pdf
2. Evertsz, C. J. G and Mandelbrot, B. "*Multifractal Measures*", in "Chaos and Fractals, New Frontiers of Science", Springer-Verlag (1992).
3. https://www.sciencedirect.com/science/article/abs/pii/0920563287900363

4. https://arxiv.org/pdf/1606.02957.pdf

5. Christensen, K. and Moloney, N. R., "Complexity and Criticality", Imperial College Press 2005.

6. Available at the following site:
   http://www.smmlab.it/zapperi/reprints/Vespignani_Physical%20ReviewE57_6345_1998.pdf

7. Available at the following site:
   https://www.researchgate.net/publication/340514250_Quantum_Field_Theory_as_Manifestation_of_S
   elf-Organized_Criticality

8. http://www.scholarpedia.org/article/Lagrangian_formalism_for_fields

9. https://arxiv.org/pdf/cond-mat/9903270.pdf

10. https://arxiv.org/pdf/cond-mat/9811365.pdf

11. Available at the following sites:
http://www.aracneeditrice.it/aracneweb/index.php/pubblicazione.html?item=9788854889972
https://www.researchgate.net/publication/278849474_Introduction_to_Fractional_Field_Theory_consolid
   ated_version

12. Available at the following site:
   https://www.researchgate.net/publication/228625050_Chaotic_Dynamics_of_the_Renormalization_Gr
   oup_Flow_and_Standard_Model_Parameters_Intl

13. https://arxiv.org/pdf/1310.5527.pdf

14. Available at the following site:
https://www.academia.edu/38752380/Multifractal_Geometry_and_Standard_Model_Symmetries

15. https://arxiv.org/pdf/1612.00462.pdf

16. https://arxiv.org/pdf/0806.2675.pdf

17. Cheng, T. P, and Li L.F., "Gauge Theory of Elementary Particle Physics", Clarendon Press, Oxford,
    1989.

18. http://maths.dur.ac.uk/users/kasper.peeters/pdf/eft.pdf

19. https://www.prespacetime.com/index.php/pst/article/view/1619/1546

20. https://www.prespacetime.com/index.php/pst/article/view/1653/1575

21. https://www.prespacetime.com/index.php/pst/article/view/1648/1569

22. https://www.prespacetime.com/index.php/pst/article/view/1656/1570

ISSN: 2153-8301   Prespacetime Journal   www.prespacetime.com
Published by QuantumDream, Inc.

Article

Faster-Than-Light Anomalies in Prespacetime & Interstellar Translocation through Hyperdimension

Chris H. Hardy[*]

This article argues that the speed of light limit C is a law applying solely to the matter and spacetime region. The infinite spiral staircase theory ("ISST") posits that prespacetime is a triune hyperdimension in which the spacetime and EM laws do not apply, because mass, space, and time are not existing yet. Yet the universe at Planck scale has an astounding non-matter energy, and the inflation and the entanglement have been demonstrated to be superluminal and/or nonlocal. Similarly, research data shows psi to be nonlocal and contradicting spacetime laws. ISST postulates that the energy filling the hyperdimension, called *syg-energy* and driven by tachyonic *sygons*, is linked to cosmic consciousness (or syg-HD), itself enmeshed with hyperspace and hypertime, and that psi and synchronicities are hyperdimensional processes. A thought-experiment is proposed that could instantiate a translocation of either information or a matter system.

Keyword: Faster than light, hyperdimension, entanglement, prespacetime, nonlocality, synchronicity, translocation.

Introduction

The first two decades of the 21st century have seen a major shift in our scientific paradigm with the discovery of dark energy and dark matter in 1998. The observation of supernovae with the Hubble Space Telescope highlighted the acceleration of the expansion of the universe and led cosmologists to the obvious inference that it had to be produced by a totally unknown gigantic force, that they called *dark energy*. And thus we had to adjust to the fact that this dark energy makes up about 69% of all energy in the universe, as goes the latest count from the Planck probe in 2015 – See https://en.wikipedia.org/wiki/Planck_(spacecraft) , and our good old (aka ordinary) matter only about 5%.

As ordinary matter is everything we have always identified as matter, from particles to stars to galaxies, it suddenly robbed the materialist paradigm (that deemed only matter to be real) of any substance, literally so, and therefore of its credibility. Dark energy is a negative energy (repulsive radiation pressure), that counteracts exactly the positive and attractive energy of gravity. As for dark matter, a positive energy, it accounts for 26% of the total energy. The nature of both dark energy and dark matter are unknown. Thus, given that 95% of the energy of the universe is a total mystery... we can assume that actual physics deals only with 5% of the universe's reality.

---

[*]Chris Hardy, Ph.D., Eco-Mind Systems Science, Seguret, France. Email: chris.saya@gmail.com

What scientists are looking for with dark energy is a repulsive energy that has to be in such an amount as to overwhelm the gravity attraction of all dark matter and normal matter combined, and it has to make up 69% of the total energy of the universe. An article on the NASA website called *Dark energy, dark matter* (See http://science.nasa.gov/astrophysics/focus-areas/what-is-dark-energy) gives us three actual candidate explanations for dark energy. The third – and the only one without major shortcomings – is "a new kind of dynamical energy fluid or field, something that fills all of space," and has been called *quintessence*. "If quintessence is the answer, we still don't know what it is like, what it interacts with, or why it exists. So the mystery continues."

1. The Sub-Planckian Region Is The Prespacetime Hyperdimension

It is only after the Planck time, when the universe was an infinitesimal fraction of a second old (in effect, 5.3×10^{-43} second), and its diameter was Planck length (1.6×10^{-33} centimeter), that the first energy-particles could be born and acquire mass in the Higgs field. Then, space, time, mass, and matter – and therefore electromagnetic (EM) fields and causality – are born, and the particles of the Standard Model will appear in due order. Thus, it's only after and above Planck scale that the matter universe will deploy itself, from the Higgs bosons to the first stars and proto-galaxies some 420-500 million years after the origin.

Physicists infer that before Planck scale, there were no particles and therefore no matter (nor was there any electro-magnetic force, strong force, weak force, but solely gravity), and that no time and no space had formed yet – hence its name of 'pre-spacetime'. And nevertheless the primordial universe had a gigantic energy, despite the fact that we are still, at Planck scale, an immense time before the Big Bang now identified with the inflation phase at 10^{-36} second. As it is, the new frontier of physics is now set on the pre-spacetime region, with several leading physicists expecting the laws governing the sub-Planckian region (or, in Sarfatti's terms *subquantum physics*) to be widely different from the matter and spacetime region.

ISST posits that below Planck scale starts the hyperdimension, a non-matter region of the universe filled with a boundless *non-matter energy*, in fact a consciousness and active information energy called *syg-energy* (i.e. *sygons* waves). But let's realize that this hyperdimension (sub-Planckian or sub-quantum region) happens both *before* Planck scale at the origin of our universe-bubble, and *below* Planck scale at each point of our 4D matter universe.

1.1. The birth of spacetime, particles, mass, & the speed of light

To get back to the quantum and spacetime region: starting at Planck's scale, quantum mechanics, string theory, and Relativity theory allow us to model the blitz expansion of the universe. Below are the presently agreed-upon events happening in the point-universe (although some tiny details may vary among scientists).

- *Planck scale* happens at 5.3×10^{-43} second.

- Between 10^{-36} and 10^{-32} second happens the *Inflation* (now assimilated to the Big Bang), an explosive expansion of the point-universe.

- Between 10^{-10} and 10^{-4} second, the *Higgs field* is formed, a super hot quark-gluon plasma, and the first energy particles – the Higgs bosons – appear within it; the temperature is down, but it still reaches one million of a billion degrees. In this Higgs field are glued bosons, quarks, gluons, electrons, photons, neutrinos. The experimental discovery of the Higgs boson, on March 14, 2013, proved that masss is not an intrinsic property but just an effect of particles interacting with the Higgs field.

- Within the first second, the *neutrinos* free themselves, they *decouple*, from the super hot plasma, when its temperature was down to 10 billion Kelvin, and are set on a course, thus creating their own 'cosmic neutrino background' (discovered in 2008, from the analysis of the photons' Cosmic Microwave Background).

- At 10^{-4} second, the massless gluons exert a *strong force* on the *up* and *down* quarks (the first and lightest quarks) that then form protons and then neutrons, each having three quarks in a sort of unbreakable connection. Thus protons, neutrons, up and down quarks, and electrons are all that is needed to form all the existing atoms.

- At 10^{2} seconds (1.40 minute) the radiation era ends and *the matter era begins*; all protons and neutrons of our actual universe have already been created. Simultaneously, the *photons decoupling* happens: the photons, which up to now had been glued to the plasma and kept popping up from the field or ocean of Higgs, only to be smashed and reabsorbed, are suddenly able, with a now colder plasma around 3000 degrees, to thrust themselves free and start on a cosmic journey that will light up the universe. It is its relic (or afterglow) image that we detect now at about 370,000 years after the Big Bang, and that forms the Cosmic Microwave Background (CMB). The photons, set on a course, will slowly get cooler (they are now at 2.7 degrees above the absolute zero) and their frequency will progressively decrease (while their wavelength increases) from the gamma rays of the origin, to the low microwave frequency they exhibit now. (According to Planck's relation $E = hv$, the energy of a photon, E, known as photon energy, is proportional to its frequency, v; and the wavelength is inversely proportional to energy.) My view is that these photons, through their travel within a sygons-filled medium, are themselves creating the spacetime as an enclosed bubble as they dart along.

- Starting at minute 3 until minute 20, protons and neutrons will combine via a process of nuclear fusion, thus forming the nuclei of atoms (a process called *nucleosynthesis*). Then the first simple atoms will form when electrons start bonding with these nuclei and orbiting them, mostly atoms of hydrogen (the lightest atoms, comprising one proton, and thus of atomic weight 1), deuterium, and helium-4. It is breathtaking to realize that the nucleosynthesis took only 17 minutes and that all hydrogen still existing at our present time in our whole universe has been formed within twenty minutes after the Big Bang.

1.2. The speed of light limit is tied and restricted to the spacetime region

We can safely say that before the birth of spacetime around Planck time, and before the onset of the matter era (at 1.40 minute) and the electrons bonding with the nuclei to form the first hydrogen atoms (at minute 3 to 20), there cannot be any EM and Relativity laws governing spacetime; and neither can the equivalence $E=Mc^2$ exist before the onset of mass, with the emergence of massive particles having crossed the Higgs field (along the famous Einstein equation that equates energy (E) to the mass (M) multiplied by the speed of light (C) squared).

Yet, there are also massless (or near-massless) particles such as photons and some types of neutrinos that can reach very high levels of energy. And we know that when the universe was the size of the first quantum (at Planck scale), it had an immense energy (10^{19} GeV – giga or million electron-volts) and a bewilderingly high frequency (1.85×10^{43} Hertz). And it seems similarly that the speed of light C, setting the limit of speed for the whole electromagnetic spectrum (from microwaves to gamma rays) is definitely bound to the spacetime region, at least from the time of the photons decoupling onward.

Spacetime, then, is literally born, either at Planck scale or else just after with the Higgs field and the first particles gaining some mass. Then all the EM radiations will abide by the speed of light C (at 3.00×10^8 m/s in the vacuum), and all speed in spacetime (theoretically at least) will be constrained by the C limit. Similarly, the EM laws, such as the inverse square-law showing a decrease of the signal/force with distance, are strictly bound to the spacetime region.

In brief, the speed of light limit – that no energy ray can exceed C – is a law applying solely to the spacetime region or 4D-matter region, namely to the EM energy spectrum, and to the quantum region. Indeed, in ISST, it is called the QST or quantum-spacetime region, because the theory predicates that the particles/strings, in their particle aspect, do abide by spacetime and the C limit (their position parameter), whereas in their weird wave and superposition aspect they belong to the hyperdimension.

In summary, the pre-spacetime region (sub-Planckian and sub-quantum) is a hyperdimensional region in which the laws of spacetime (EM laws and the speed of light limit) don't apply, because matter (particles), mass, space, time, and material causality are not existing yet. However, at Planck scale, the point-universe has already an astounding energy, as we mentioned, calculated to correspond to 10^{19} giga electron-volts, and we can only assume this energy to be widely accrued the closer to the origin. If matter, mass, and the C limit don't exist before Planck scale, and yet the point-universe has an immense energy, then this energy is by definition of a non-matter type, and linked neither to mass nor to C. In effect, as posited by ISST, it is a hyperdimensional energy, its virtual speed vastly exceeding C and this syg-energy that fills the hyperdimension is essentially (or rather, "quintessentially") linked to cosmic consciousness.

2. Light Speed Anomalies & nonlocal processes

2.1. Inflation phase deemed at billion times *C*

However, there is a first major anomaly in terms of EM laws and the speed of light happening above Planck scale (about 10^{-43} second) and before the Higgs field starts at 10^{-10} second. In effect, during the inflation phase between 10^{-36} and 10^{-32} second (an infinitesemal fraction of a second after Planck scale), the point-universe shows an explosive expansion and its size will be multiplied 10^{50} times. Yet, the size of the point-universe is so small, and the forces it houses so huge, that it stretches our imagination to try to figure its reality. The Big Bang is now considered to be this inflation phase, an immense time after the X-point of origin.

The creators of Inflation Theory, first Alan Guth and then Andrei Linde, estimated the speed at which this happened to be *billion times that of light*. The inflation model had been put into question by some physicists opposed to a universe having such origin; but it is indirectly supported by the high coherency, despite a faint but crucial anisotropy, found in the cosmic microwave background.

As shown on the time axis of Fig. 1 (in seconds at the top), the present cosmological understanding is to view the onset of the *matter era* precisely during the inflation. Yet, in ISST framework, given that the inflation happens so early after the X-point of origin and well before the Higgs field and its plasma of particles, we cannot dismiss the possibility that this enormous speed would reflect a hyperdimensional energy still operating – in which case the spacetime would really start later with the Higgs field, with the onset of massive particles exiting it. And since the neutrinos had decoupled from this ultra hot Higgs plasma by the first second, we can consider that their speed was still immensely greater than that of the inflation phase, and therefore of C.

In this perspective, the neutrinos would be the last HD sygons (with the lowest HD frequency) to cross Planck scale and the (just forming) Higgs field without acquiring mass. ISST assumes that the Higgs field is created by the low-frequency sygons (issued from the late-time and larger turns of the ISS spiral of the origin) whose wavelengths, at one point, became so large as to start interfering, thus creating a turbulent foam quickly densifying, through which the subsequent and even larger sygons will have to pass, as whirling loops, acquiring mass in the process; the massive particles exiting the Higgs field will retain the HD sygons at their core, as a compact sub-Planckian 5th dimension.

Also, the *domain of torsion waves* has seen a volcanic development in Russia lately, with physicists such as Shipov and Akimov. According to them, torsion waves can connect systems at an immense distance at a speed that is a billion times that of light ($10^9\ c$). The interference patterns of the torsion waves of stars and solar systems would create a galactic hologram, and those of galaxies would create the universe hologram, through which all galaxies and stars can evolve coherently.

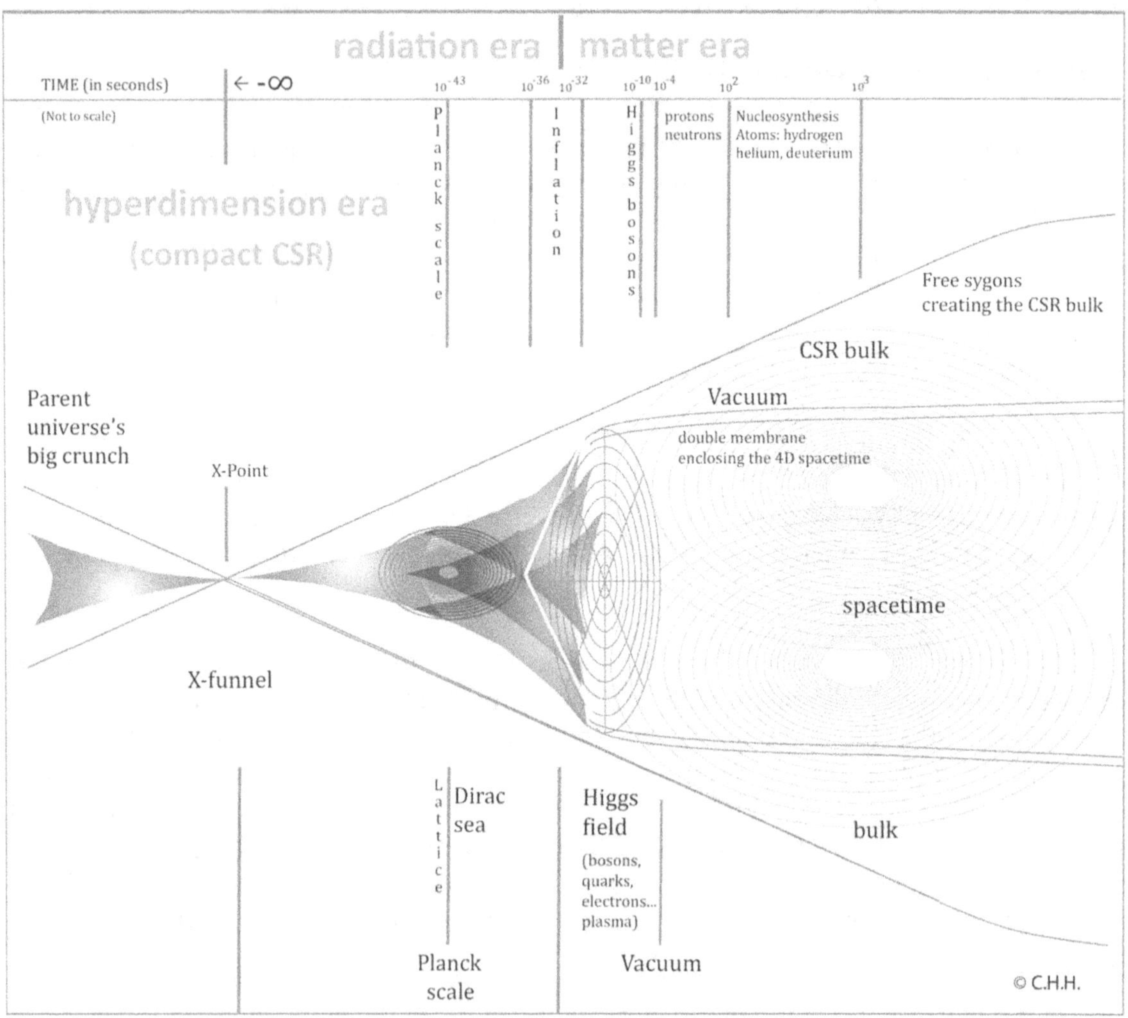

Fig. 1. Timeline in the Infinite Spiral Staircase theory. From the X-point of origin: sygons creating both the CSR-HD bulk and spacetime. Concept and digital artwork by Chris H. Hardy. (Cosmic DNA, 313)

2.2. Entanglement: Communication beyond spacetime

As I have argued in a 2017 paper called "*Nonlocal consciousness in the universe*" there are five types of processes that display a 'beyond spacetime' property or nonlocality. Not only are these at odds with Relativity and EM laws, but also with the indeterminacy of quantum mechanics. We already saw the first three: with the first one being the sub-Planckian region (both the pre-spacetime at the origin, and the subquantum region at any coordinates of spacetime; the second the beyond-spacetime, non-matter, dark energy; and the third the faster-than-light speed during the inflation phase. As for the last two nonlocal processes, the fourth is the entanglement and the fifth type is psi phenomena such as telepathy – proven to operate beyond-brain and beyond-spacetime. Of these five nonlocal (beyond-spacetime) processes, only three present a clear and demonstrated beyond-C (or tachyonic) anomaly; yet, in ISST, I argue that psi, telepathy, and synchronicities are hyperdimensional processes that operate via the syg-energy of the HD, with virtual speed vastly superior to C (Hardy 2015, 2017).

Let us now focus on the entanglement. In fact, nonlocality has been established via the entanglement experiments, whose protocol has been devised by John Bell, based on the famous Einstein-Podolsky-Rosen, or EPR, thought experiment. According to a series of experiments conducted by Alain Aspect between 1982 and 1984 that conclusively proved the entanglement, particles issued from a unique source (i.e. they are "paired") abide by Pauli's law of spin, in the sense that their spins must remain complementary. He posited that two electrons can share the same orbital at the condition that their spins be opposite, meaning that the spin of such paired particles can only take the values, for example, -1/2 and + 1/2. For instance, if the spin of particle A is modified (as with a mirror), then particle B will immediately change its spin to remain complementary.

One experiment implied one of the particles being bounced on the moon, and the enormous distance thus gained between the two paired particles showed without a doubt that the exchange of information (or nonlocal correlation) between the distant particles couldn't imply a signal transmission through space, given that the correlated change of spin happened quicker than the speed of light. Citing a 2012 paper by John Matson, the Wikipedia article on *Entanglement* states (https://en.wikipedia.org/wiki/Quantum_entanglement) "so-called "loophole-free" Bell tests have been performed in which the locations were separated such that communications at the speed of light would have taken longer – in one case 10,000 times longer – than the interval between the measurements." Thus, we have an experimental proof of an exchange of information (at the very least) 10,000 times C.

Since space is indissolubly enmeshed with time in the spacetime of Relativity, the entanglement reveals a 'beyond spacetime' process. Thus, while the paired particles (photons or electrons) are themselves existing in spacetime (in their particle aspect), *their entanglement instantiates nonlocality* (in their wave aspect). Furthermore, as I've proposed it in ISST, the wave component and wave processes of particles belong to the hyperdimension, and thus their connection and entanglement is, in effect, an instantaneous exchange by the Rhythm hyperdimension (hypertime in ISST).

Hu & Wu (2013) postulate that "quantum entanglement arises from the primordial self-referential spin processes which are envisioned by us as the driving force behind quantum mechanics, spacetime dynamics and consciousness." Moreover, they posit these spin processes as instantiated "in non-spatial and non-temporal prespacetime" – a predicate that is in accord with ISST in more than one respect. Firstly because in their view and that of ISST pre-spacetime and spin are associated with consciousness. Secondly because spin is also an essential dynamics at the origin in ISST.

In fact, the *Infinite Spiral Staircase* (the ISS) of the origin is a golden spiral, therefore constituted of quarter-circles (the staircase steps) whose radii follow the Fibonacci sequence. This spiral issued from the White Hole of the origin with a near-infinite energy is itself spinning outward, but moreover, each quarter-circle is a virtual circular closed-string (the full circle itself with this radius and frequency); the whole ISS spiral is thus a near-infinite set of primordial spins that, beyond their angular momentum, are also each one associated with a specific frequency (and its harmonics).

Then, each quarter-circle launches a torsion wave (a sygon) of this frequency, and this closed string is also spinning, and these sygons will pervade the (future) matter universe, some being at the core of all particles as a 5th dimension, and others crisscrossing the universe as Free Sygons. And thirdly, while it is an agreed-upon fact that the opposite spin of fermions (asymmetrical) particles is what instantiates the entanglement, we still have, with particles, one foot in the post-Planck region (QST) of matter, C, and mass. But if, as ISST hypothesizes, the wave component of the particles, and therefore their nonlocal entanglement, is a dynamics pertaining to the hyperdimension – both the Rhythm-Rotation and the Center-Circle HD, these two being enmeshed with hyperconsciousness –, then it fully corroborates Hu and Wu's postulate.

In summary, there are three major demonstrated anomalies regarding the speed of light limit: (1) the sub-Planckian region (which exists in pre-spacetime, in each particle, and also at any point of spacetime); (2) the inflation phase; (3) the entanglement. All of them, with the addition of a fourth one, psi phenomena, are hypothetized in ISST as hyperdimensional processes driven by the tachyonic syg-energy of the hyperdimension.

2.3. Some astronomers deem C & gravity originally 10-20 billion times faster

A periodicity has been discovered in the redshift of galaxies, called redshift periodicity or quantization, by William Tifft studying the Coma cluster and corroborated by several astrophysicists in the 1990s, such as Bruce Guthrie and William Napier (from their 1997 study of 250 galaxies). It means that the redshifts of cosmologically distant objects (in particular galaxies) tend to cluster *around multiples of some particular value*; and furthermore, it implied that *the speed of light had been enormously higher at the origin*. In 1987, V. S. Troitskii had argued that the lightspeed had originally been about 10^{10} times faster than now. Then several mainstream physicists argued and published that the lightspeed was hugely higher at the Big Bang scale (such as Joao Magueijo of the Imperial College in London, John Barrow of Cambridge, Andy Albrecht of the University of California at Davis, John Moffat of the University of Toronto).

All this led some astrophysicists such as Tom Van Flandern and Tifft to hypothesize a *wave nature* to the interstellar void, with a higher density of stellar clouds on the force lines. Van Flandern then put out a theory of gravity (based on Le Sage work) in which G derives from "a flux of invisible 'ultra-mundane corpuscles' impinging on all objects from all directions at superluminal speeds," these particles offering limitless free energy. In a 1998 paper, he argued that *gravity (the graviton), based on observations, is "not less than" twenty billion times the speed of light.*[1]

Van Flandern bases his argument on astronomical observations, the crucial one being that all astronomers were taught "that all gravitational interactions between bodies in all dynamical systems had to be taken as instantaneous, [… and] to calculate orbits using instantaneous forces," since introducing a lightspeed delay into gravitational interactions would give results disagreeing

---

[1]Van Flandern, T (1998). "The speed of gravity ? What the experiments say". Physics Letters A. 250 (1–3): 1–11. Full text at http://www.ldolphin.org/vanFlandern/gravityspeed.html. See also the full text of his 2003 article on "Allais gravity…" https://www.researchgate.net/publication/228453035_Allais_gravity.

with observations. (These interactions are distinct from gravitational radiation and waves, which propagate at lightspeed.) And in fact, says he, "It is widely accepted, even if less widely known, that the speed of gravity in Newton's Universal Law is unconditionally infinite." Whereas in General Relativity (GR), gravity is an effect of the geometry of curved spacetime and of mass affecting this geometry, and is not considered a force that propagates. Yet Van Flandern underlines that a number of problems remain with GR: "How the external fields between binary black holes manage to continually update without benefit of communication with the masses hidden behind event horizons. (…) Why do total eclipses of the Sun by the Moon reach maximum eclipse about 40 seconds before the Sun and Moon's gravitational forces align? How do binary pulsars anticipate each other's future position, velocity, and acceleration faster than the light time between them would allow? How can black holes have gravity when nothing can get out because escape speed is greater than the speed of light?"

3. The Infinite Spiral Staircase Theory

ISST posits a triune hyperdimension composed of three interwoven strands: hyperspace, hyperconsciousness, and hypertime (See the 2015 book *Cosmic DNA at the Origin* and several articles since 2015[2]). This triune HD sprang forth from the origin (a White Hole coupled with the Terminal Black Hole of the parent universe) as a golden spiral bearing on the quarter-circles composing it a near-infinite data bank of frequencies (the steps of the "staircase") based on the logarithm of Phi (a quasi-Fibonacci sequence). This data bank was (and is still) bearing a near-infinite active information field about all beings and systems optimized in parents universes (thus acting as a cosmic and hyperdimensional DNA) as well as about all new systems existing and evolving in our universe. Each quarter-circle of the ISS sent a torsion wave of the specific frequency and its harmonics, in effect a virtual closed-string (or closed-brane) of immensely faster-than-light (FTL) speed and spin (rotation), called a *sygon*, along the drift of the spiral, in a large cone.

The highest frequency sygons, nearer to the origin, called the Free Sygons, formed the bulk of the HD (see Fig. 1 above); then the low-frequency HD sygons (with larger and larger wavelengths) started interfering at the mouth of the ISS spiral at Planck length, thus forming the quickly densifying and turbulent Higgs field (as an orthogonal near-flat round surface) – a field that the subsequent whirling sygons now had to cross (precisely, to spin and tunnel through), thus acquiring mass and a cloud of charge, and leading eventually to the onset of the particles of the Standard Model and the matter era. Each particle and atom will thus keep, at its core, the original HD sygon, thus forming its 5th dimension, curled up and compact at a sub-Plankian scale.

ISST postulates that the massive particles exiting the Higgs field started the Quantum-Spacetime region (QST) as a near-cylinder enclosed within the larger and already existing HD bulk. One specific triple prediction of the ISST is that, as a consequence of the rotational momentum of the cosmic ISS of the origin, and its enormous negative radiation pressure, as well as that of the torsion waves of all sygons, *the Higgs field* (1) was itself set in rotation, and it extended and

---

[2] Full articles on https://independent.academia.edu/ChrisHHardy/Papers.

curled itself around the whole QST region, (2) thus imprinting its curvature on spacetime, and (3) it became the *curved* vacuum membrane, turbulent with the HD-sygons creating loops and bubbles of polarized vortices while crossing it back and forth (Hardy, 2015b, "The quantum vacuum…")

3.1. ISST: The triune CSR hyperdimension & the HD sygons

The three strands or 'dimensions' of the hyperdimension are enmeshed, just like our 3D of space are enmeshed with 1D of time in our 4D spacetime region:

- *Center-Circle HD,* or **hyperspace**, allows the *self-organization* of individual systems, that is, a creative, innovative, and negentropic force which, enmeshed with the syg-HD, creates ever novel organization and underlies biological evolution. In the ISS, the golden spiral at the origin, Center-HD is governed by Pi.

- *Syg-HD* or **hyperconsciousness**, is a hyperdimensional layer of consciousness in all beings and systems in the universe, called semantic field or syg-field in short. Thus, as human beings, our HD layer, or personal syg-field, is our Self (soul, atman). Syg-HD, at the global level (the HD bulk), is cosmic consciousness – as the ensemble of all the HD-Selfs and as the Whole or the One field. This HD instantiates a permanent creation of meaning and information, of collective intelligence and knowledge in the universe, as well as spontaneous connections based on meaning and sympathy (such as synchronicities and psi).

- *Rhythm-Rotation HD* or **hypertime**, operates via *rhythm and resonance*, and allows an instant communication at great distances between systems. In the ISS – the golden spiral at the origin – Rhythm-HD is governed by *Phi* (the golden ratio) and the logarithm of *Phi* applied to the radius of the expanding spiral.

The transition from the matter region to the HDL region (and vice-versa) happens in the singularities of black and white holes. Thus, in the parent's universe's terminal black hole (BH), all matter is translated into hyperdimensional (HDl) syg-fields and these complex syg-energy fields will be retranscribed into bio-systems and matter-systems after exiting the White hole of the origin. Thus, in between universe bubbles – precisely in between the preceding universe's terminal BH singularity, and the origin's white hole singularity (the orthogonal membrane) –, there is a region of pure HDl syg-energy.

Interestingly, the co-author of a recent study of stars orbiting our supermassive black hole *Sagittarius A** at the center of the Milky Way, Andrea Ghez of UCLA, states they have corroborated Einstein's General Relativity view of gravity: "In Newton's version of gravity, space and time are separate, and do not co-mingle; under Einstein, they get completely co-mingled near a black hole." (Do, Hees, Ghez, 2019). In ISST's viewpoint, a BH's singularity opens on the HD, and in the CSR-HD, hypertime and hyperspace are indeed co-mingled and interlaced with hyperconsciousness.

Here is the fundamental principle regarding the CSR-HD in ISST:

- All matter systems, at all scales, from particles to galaxies, have an hyperdimensional layer, including ourselves. The HD layer of organic and ecological systems, and even matter systems, is a proto-consciousness called an eco-field, thus instantiating a type of panpsychism (Hardy, 2017).

- As for our global and personal HD layer, it is our own syg-field or Self (spirit, atman, soul); and the HD layer of our body is our body-consciousness. This HD Self is the immortal being (soul) of the incarnated person, and it is a fully autonomous self-conscious, and in fact supraconscious, entity dwelling in the CSR-HD. It is thus in constant communication (1) with the cosmic ISS (bearing the Akashic information), (2) with resonant syg-fields (minds, ideas, places, objects, etc.).

- Even our matter universe has its global HD layer: the cosmic consciousness (brahman, Tao), Jung's collective unconscious; in ISST, the CSR hyperdimension.

3.2. ISST: 3 regions in the pluriverse

In the ISST cosmology framework, the pluriverse is composed of three regions:

1. ***The Quantum-Spacetime (or QST) region***, in which three forces operate – electromagnetic force, strong force, and weak force – and in which the EM laws and the C limit hold sway. Spacetime is a curved space, as in Relativity theory, but in ISST, in contrast with current models, the quantum vacuum is a curved membrane enclosing it and making the boundary with the HD bulk.

2. ***The CSR Hyperdimension region***, in which hyperconsciousness (mostly the HD Selfs of individuals), hyperspace, and hypertime operate conjointly via syg-energy and a spontaneous connective dynamics based on meaning and fine-tuned by three main parameters – semantic proximity, semantic intensity, and coherence (C. Hardy 1998, 164-6). Intuitive intelligence, psi capacities, and high states of consciousness belong to this CSR-HD and exhibit the constant influence of consciousness and individualized Selfs on their environment and the matter systems they interact with.

   This HD layer both on its cosmic ISS and in all systems' individual ISS (as a 5th sub-Planckian dimension) holds the (ongoing, evolving) information on all beings and systems in the universe, thus acting as a complex Akashic field of active and self-organized information, keeping the memory of all past and current states of systems, as well as their probable future paths. The individual ISSs keep a two-way exchange of information about the state of their system with the cosmic ISS, and this is how the latter is imprinted in realtime about all changes in all systems in the universe (and why the vacuum menbrane is crossed both ways). The syg-energy units of the CSR-HD are the sygons (that constitute syg-energy); they are torsion waves made of circular/annular virtual strings at the high-frequency range of the sub-Planckian cosmic ISS.

3.   ***A quantum vacuum membrane*** surrounds the QST region (of positive energy) and stands as its boundary with the CSR-HD (dark energy, negative). As mentioned above, the Higgs field at one point becomes the vacuum's Zero Point Fluctuations field (virtual particles with maximal quantum turbulence). ISST is in full agreement with this vacuum seen as a brane (surface), modelled by Jack Sarfatti (2006) as a complex two-dimensional boundary separating the spacetime from the Dirac sea or dark energy (negative energy), in which all anti-particles (antimatter) reside; for Sarfatti, these are linked through wormholes to their paired matter particles in spacetime and form "stable bound pairs," lots of them being network-linked, thus creating "loop vortices" and "networks of loops."

ISST adds the prediction that this quantum brane is a curved surface, which would make the integration easier with the curved spacetime – modelled by Einstein on Riemann's hypersphere –, while remaining consistent with quantum theory, which, both in the Standard Model and the superstring M-Theory, is set in a flat space. Let's note that in his 1854 thesis presentation, ten years before Maxwell, Bernhard Riemann, a German mathematician, was the first to predicate higher n-dimensional space – such as a 4th dimension of space in order to model a hypersphere at 3 dimensions, thus inventing the maths of the curved space that Einstein elaborated on.

In this paper Riemann stresses that if we ascribe to space a constant curvature, not only is it an unbounded surface, but this surface, "in a flat manifold of 3 dimensions would take the form of a sphere, and consequently be finite." (3.2. p. 660). And also, concerning the infinitely small: if we take the metric of space to be a continuous manifold, its "ground must come from outside, (…) in binding forces which act upon it. (3.3. p. 661). And ISST is in accord with that in two ways: the QST bubble is finite within the CSR-HD, and it gets its Einsteinian curvature from the immense negative radiation pressure of the CSR-HD. Also, QST being finite, it calls for both a beginning and a big crunch of the matter universe bubble (the Terminal Black Hole).

3.3. Syg-energy: Two types of hyperdimensional FTL sygons

Altogether, the *Center-Syg-Rhythm HD* (CSR-HD) is filled with a hyperdimensional (HDl) energy called *syg-energy* (whose virtual particles-waves are the sygons), that both pervades the HD and acts as its core dynamics. Syg-energy is a connective dynamics driven by meaning/semantics (Syg-HD), that is self-organizing (Center-HD), and that instantiates HDl frequencies (Rhythm-HD), all at once.

Some cosmology models use both a 'compact, sub-Planckian HD' (as in Kaluza-Klein theory) and a 'bulk HD' (as in the 1999 Randall-Sundrum theory), such as Bernard Carr's universal *higher dimensional information space* (Carr 2003, 2009), and my own ISS Theory. *So that in ISST, the HD sygons are of two types: the Free Sygons* of very-high-frequency (VHF) that now fill the bulk of the HD and compose our HD-Self; and the low-frequency sygons at the core of the particles and all matter systems (as a 5th dimension), whose collective field makes our body-consciousness ("low-frequency" in terms of the HD, thus at least above, and as a multiple, of Planck frequency).

4. Beyond Space Communication & Travel via Hyperdimension

4.1. Speed of communication in and via the HD

The VHF Free Sygons crisscross the universe unimpeded at billion of billion times the speed of light, as torsion waves that spontaneously link resonant, sympathic, or coupled systems, for example two friends bonded by love and empathy, sharing a telepathic information, or a person having a sudden clairvoyant information about burglars trying to penetrate into their beloved country house several hundred miles away (as it happened to me).

I hypothesize that the virtual speed of communication and exchange in the HD is an enormous multiple of C calculated to be proportional to the frequency of torsion waves sygons – and thus tends to the infinite near the origin. In all practical purposes, the VHF sygons that created the bulk of the HD (some clusters of which constitute our immortal Selfs and the information or syg-fields of all systems) instantiate a near-instant communication, whatever the distance, as shown in the entanglement experiments of the EPR type. As I suggested earlier, ISST postulates that, as any matter system made of particles, Shrödinger's Psi equation would have its own HD, in its wave and superposition component.

As for the sygons at the core of particles and all matter and massive systems, which, as all sygons do, keep interacting and exchanging information between them, while of a lower HD frequency, their velocity is still hyperdimensional and widely higher than C, and they could be a good candidate for instantiating the instantaneous attraction between celestial and stellar masses, in other words, for gravity as a force.

Given that ISST posits that the highest mental, intuitive, and spiritual processes are instantiated by the syg-energy of our (personal) hyperdimensional Selfs, these thought processes partake of the properties of syg-energy and notably of FTL speed. Since a form of synthetic thought (non tied to language) is used in telepathy, I predicate that psi communication involves immensely faster-than-light speed. So that telepathy with alien beings residing on exo-planets in our galaxy does seem instantaneous. As for an astral travel (with the lower-frequency sygons of our energy-body), it appears instantaneous to anywhere on Earth (as goes my experience); and only if the world is far beyond our solar system, does it feel like it takes some short time.

4.2. The CSR hyperdimension as annulling space in the entanglement

The HD Rhythm-Rotation shows all quantum-scale systems (particles/waves) have some spin, and that some may present a double rotation in a torus – both clockwise and anti-clockwise, such as the whole hourglass system of two entangled particles we will now explore. This HD allows information to be shared instantaneously between two systems that are distant in Einsteinian space but that have the same frequency or Rhythm. But let's remember that this information is not an abstract one, but one filled with meaning, intention, and generally the attributes of consciousness (aka of the syg-HD). Rhythm-HD instantiates a real in-formation: a spontaneous communication and influence through the rhythms and the specific frequency (or network of frequencies) of the systems in rotation.

- Whatever their scale, any rotating system, or rhythmic form (such as a spiral shell) – sygons, electrons in the shells orbiting an atom, or planets orbiting a star – all have a signature in the Rhythm-HD. Thus, the electrons' orbits are a complex and plural Rhythm-system with both types of spin – angular momentum (rotation/spin as up, down or zero) and orbital angular momentum; similarly, each planetary system (with two basic rhythms for each planet, rotation and orbital revolution) shows a complex rhythm consisting of all planetary rhythms.

- To the Rhythm is attached the parameter of the positive or negative spin and its value as integers or half-integers (also referred to as spin-up, spin-down, and spin 0 (ground state).

- Each spinning sygon, particle, or system has its own frequency (or network of them) and energy as function of this frequency (according to Planck's relation).

- Rhythm-HD is a nonlocal communication, as all resonant or proximate rhythms communicate between them, and all identical but exactly inverted rhythms are entangled. (In fact, it's the opposition of their spin/rotation that is maintaining the system in an entangled state, and the entire hourglass within the Syg-manifold).

- In brief, Rhythm-HD is annulling the 3D space.

Thus, Rhythm-HD allows us to model the EPR paradox in a new light:

The entangled system with two paired particles with opposite spins can be represented, in Rhythm-HD as a unique rhythmical system – as a hourglass with 2 coupled spinning tops attached at their small end, and turning in the opposite direction:

- For a unique system (such as a closed-string sygon) the spin direction, the angular momentum (rotation speed) and the frequency are the Rhythm-HD parameters.

- For 2 entangled (asymmetric) paired particles, it is the double and inverted spin, angular momentum (rotation speed) and frequency that are the Rhythm-HD parameters.

In the entangled system, the 2 paired particles are coupled via their identical but opposite spins. Notwithstanding the ever increasing spacetime distance between them, their double Rhythm system remains a unique entity in the Rhythm-HD.

If one the 2 spins (Rhythm-HD) is mecanically or artificially inverted, the other one instantly gets inverted in order to maintain the integrity of the system; because, in Rhythm-HD, their double-rhythm is still unique and enmeshed, and they continue to respond to each other and to be coupled in the HD (beyond spacetime), whatever the parameter of distance in spacetime. It means that it is not the 2 particles per se that are entangled, but rather their hyperdimensional Rhythm as a set of parameters. And thus, to perturb (in effect, to invert) one of the 2 spins amounts to rolling over the whole hourglass.

ISSN: 2153-8301   Prespacetime Journal   www.prespacetime.com

Published by QuantumDream, Inc.

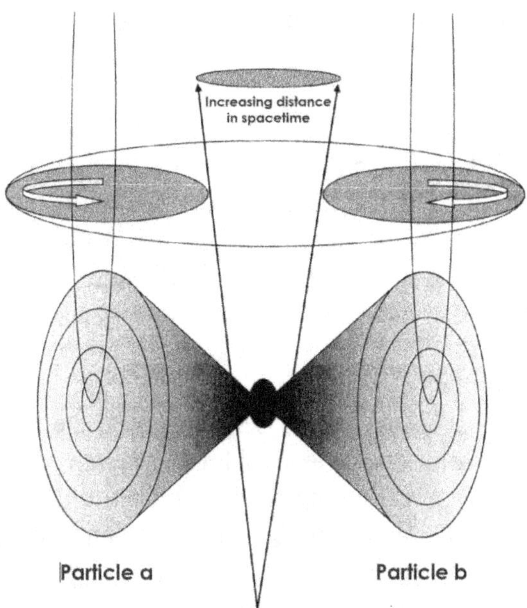

Fig. 2. Entanglement of particles in the ISS theory. The paired entangled particles constitute a complex but single system in the Center-Syg-Rhythm hyperdimension – represented as an hourglass with two spinning tops rotating in an opposite direction. They remain interconnected via their Rhythm-HD, whatever the increasing distance separating them in the 4D-spacetime (the vertical cone). (Digital artwork, Chris H. Hardy, taken from La Prédiction de Jung, 2012, 400)

This double-funnel structure is the same one as the Syg-Funnel appearing to cross the space dimension and to link two distant telepathic-harmonic fields (or Telhar fields, as recounted in *The Sacred Network*, 272-76). It is also identical to the X-Funnel at the origin of the universe (the BH-WH Kerr double system).

Thus this X-shape (double-funnel, hourglass-shaped) structure, as expressed in the Tibetan Dorje, is nothing less than an archetypal form of the CSR-HD. The quantum entanglement can allow only multiples of 2 paired particles with opposite spins. It seems that systems with four (2x2) entangled particles – a double X-funnel, or double hourglass, or double Dorje – are also such archetypal forms, as shown in the Cross, and cross-shaped Dorje.

It is to be expected that systems with more entangled particles or units do exist, on a base-2. Indeed, it was a core step in Pauli's elaboration of the Law of spin, that he posited "a new quantum theoretic property of the electron, which I called a 'two-valuedness not describable classically,'" after which "the general formulation of the exclusion principle became clear to me" (in 1925). (Pauli's Nobel lecture, 1946, 29).

4.3. Hugely Faster Than Light communication and travels

Thus we saw that:

1. Rhythm-HD shows the property of annulling 3D-space between particles entangled via their spin direction, angular momentum (rotation) and frequency; and

2. to perturb one of the 2 spins (of the paired particles) amounts to rolling over the whole hourglass in the HD.

It is evident that the entanglement of particles, at least in the last two or three decades, has been used as a fertile ground for the research into sending and receiving information at FTL speed, indeed quasi instantaneously; and that the exchange or translocation of matter is also being experimented along the same type of entangled systems protocol; this despite the fact that most of the forefront research in the world is probably classified and hushed.

The Wikipedia article on *Quantum teleportation*[3] claims that "quantum information (e.g. the exact state of an atom or photon) can be transmitted (exactly, in principle) from one location to another" (provided a previous "quantum entanglement between the sending and receiving location," but that "it cannot be used for faster-than-light transport or communication of classical bits. While it has proven possible to teleport one or more qubits of information between two (entangled) quanta, this has not yet been achieved between anything larger than molecules." It also insists that "quantum teleportation is limited to the transfer of information rather than matter itself."

The ISST perspective departs from this stated limited view. The argument in 4.2. above allows us to fathom that inverting the spin of the particle at one end of the Rhythm-HD hourglass, provokes in reality the instant translocation of particle A (with its complete system's syg-field information) to the location and momentum of particle B, while particle B takes the 4D-locus of particle A. So that it would not be solely an exchange of information between the two systems at the two ends, but rather a double and *reciprocate translocation – a swapping*. Each system takes the place of the other one, whatever the spacetime distance between them – and that means not only whatever the *4D spatial distance* between them, but also whatever the *4D time distance* between them. It thus opens the possibility of transferring or translocating instantly a system set at the A end, to the B end, and vice-versa.

However, let's not forget that in order to make use of this hyperdimensional dynamics, one of the particles/systems has to be dynamically trapped in the spacetime coordinates of the receiving/departure gate, while the other one has necessarily to be sent at or below the light speed at the gate of sending/arrival. So that, in this protocol at least, even if the 'gates' are somehow stabilized and fixated (and they would have to be for all practical purposes), to establish them in the first place would take a 'normal' spacetime travel duration at either particle speed or at the state-of-the-art interplanetary or interstellar travel speed.

---

[3] https://en.wikipedia.org/wiki/Quantum_teleportation

5. FTL Interstellar Travels

Let us now focus on the alien crafts and 'fastwalkers' speeding though our interplanetary space and observed by astronomers – such as Jacques Vallee when he was working in his youth at Paris Observatory, something he recounts in his biographical Journals *Forbidden Science* and in his sci-fi *Fastwalker*, observations that, unsurprisingly, were dutifully hushed and erased from the official books. Observations made by the cameras of diverse satellites, or the most sophisticated telescopes on Earth, or else by amateurs astronomers; for example, of crafts crossing the entire moon surface in a few seconds. Or else observations recorded on the diverse cameras and radars of Fighter jets, as making uncanny gravity-defying turns, decelerations and accelerations deemed "impossible" and that would be instantly lethal for us humans, unless the whole craft was in an anti-grav field. Or else the numerous observations of instant (or near instant) disappearances of the alien crafts, as shown on the footages the Navy pilots' intercept of a UFO/UAP, namely, three clips, released by the Navy between December 2017 and March 2018 by *To The Stars Academy of Arts & Sciences*.

These three clips of declassified military footage are "unidentified aerial phenomena [UAP]," Navy spokesperson Joe Gradisher confirmed to CNN.[4] Of course I'm not interested here in analyzing these too numerous cases, but in the principles of such interplanetary or interstellar modes of locomotion. Let's remember the boomerang-moving interstellar visitor Oumuamua, which made a large turn around our sun and approached it on September 9, 2017; its trajectory took it on a path near to Earth, with its closest approach to Earth within 15 million miles only, on October 14, 2017. Oumuamua came from the direction of the constellation Lyra, at about 57,000 mph (92,000 km/h) relative to the sun, and was observed dashing away at 97,200 mph (156,400 km/h) toward the constellation Pegasus (thus enigmatically increasing its speed by 71% in a way that couldn't be explained by the gravitational slingshot it got from orbiting the sun).[5]

And just to keep things in perspective, the Earth orbital speed around the sun is about 67,000 mph (107,000 km/h), and the sun (and our whole solar system) orbits our Milky Way galactic core at a velocity of 828,000 km/hr, and yet, while at about 28,000 light-years from this core, we take about 230 million years to complete one orbit (see fig. 3). And our sun's nearest neighboring stars in the galaxy are the triple star system of Centaurus (Proxima Centauri, and Alpha Centauri A and B, at 4.23 Ly, 4.32 Ly, 4.37 Ly respectively). Our fastest probe, New Horizons (sent to Pluto in January 2006 and that flew by Pluto in July 2015), travels at 33,000 mph. It takes light and radio waves 4.5 hours to get to Pluto.

---

[4] "The US Navy just confirmed these UFO videos are the real deal" by Scottie Andrew, CNN, 9/18/2019. (https://www.cnn.com/2019/09/18/politics/navy-confirms-ufo-videos-trnd/index.html)
[5] See the 2 posts on my blog, at: http://chris-h-hardy-dna-of-the-gods.blogspot.com/search?q=Interstellar

Fig. 3. The main arms of our Milky Way galaxy and our Sun and solar system on the Orion arm. Credit Starchild website, NASA/ GSFC.

5.1. Space-faring aliens most probably use the syg-energy of the hyperdimension

My view about aliens roaming the galaxy, expounded in my sci-fi *Space Allies*, is that very ancient alien civilizations (namely the ones integrated in our galactic Federation), have since ages discovered the boundless and tachyonic (FTL) energy of the hyperdimension (called *syg-energy* in ISST), and this is their primary source of energy (just as electricity is for us) that powers all their machines and allows them interstellar communications and travels via the HD, thus beyond spacetime. And consequently, they have no use for electricity, radio, or TV EM frequencies anymore.[6] Furthermore, syg-energy (the sygons, virtual closed strings and torsion waves) is much more than just a FTL-type of energy: let's remember that in the triune HD, hyperspace and hypertime, interlaced, make up the HD texture and are enmeshed with Syg-HD – the energy of consciousness, or rather consciousness-as-energy. So that in any sygon-based video, photo, or hologram, there's a dimension of psycho-mental depth, that allows to 'read' the emotions and thoughts of the intelligent entities without implying any real telepathic gift, and what's more, that can reach into the past.

This book explores syg-energy and envisions it has seven bands (just like radio and TV are frequency bands in the EM spectrum), and that only the first two bands have been mastered by the aliens, yet these suffice to power all machines, interstellar syg-coms, depth syg-hologram, and instant Be-SpaD travels (beyond-space-displacement) from one "sidereal node" or gate to another one. In *Space Allies*, the most advanced Ur scientists – bent on tackling the immense depth of consciousness – are starting to explore the Syg-3 band, which opens the possibility of reconstructing the local environment of any past event, anywhere, complete with its psycho-mental information, and to "step into the event' (as an alert mind with a virtual body) while being able to influence the minds of the protagonists in this event.

---

[6] Let me note that this sci-fi was an anticipation on the ISST theory, and a novel visionary take on my previous (cognitive sciences) Semantic Fields Theory (*Networks of Meaning*,1998), which already postulated syg-energy as a beyond spacetime and tachyonic energy, but still lacked its belonging to a hyperdimension.

5.2 Anti-gravity propulsion

Of course there are the anticipated anti-gravity propulsion systems filling our sci-fi visionary landscape, and they will very certainly happen and be fun on Earth (take my sci-fi writer's instinct for it). But this is certainly not the solution for interplanetary or interstellar travels, since the gravity due to Earth mass (whose gravitational field on the surface of Earth accelerates the fall of matter at +9.8 m/s^2) decreases with height, and zero gravity happens at a height of 3200 km (one half of the Earth's radius) and upward we are free from it. As for interstellar space, it has very low density and pressure, and is only an approximation of a perfect vacuum, since it contains a few hydrogen atoms per cubic meter. Gravity is the sole fundamental force (out of three) that we know is operating at a sub-quantum scale; the strength of gravity is 10^{-41} in the scale of the proton, and 10^{-36} in the scale of the quark.

And in these boundless reaches of this sub-Planckian hyperdimensional region (from Planck length back to the origin), a region devoid of any positive mass able to distort spacetime, gravity is bound to be a very different force indeed, and to operate along the HD dynamics and parameters of the CSR-HD – namely Center-HD (hyperspace), Syg-HD (hyperconsciousness), and Rhythm-HD (hypertime). Thus, General Relativity's view that gravity is solely an effect of massive bodies distorting the geometry of curved spacetime, cannot and doesn't hold in the sub-Planckian region devoid of matter and therefore of mass.

In a 2016 article called "The Birth of Gravity and Entropy and The ISS Theory of Cosmic Origin," co-authored with physicist John Brandenburg, we have argued that "Gravity can be fundamentally tied to entropy as a spectrum of states, and that this co-dependence of gravity and an entropic state-space in the Standard Model universe (post Planck scale) makes it necessary that it be founded on a sub-Planckian set of states, or frequency spectrum, thus instantiating a sub-quantum physics. This is in accordance with Hardy's ISS theory that posits a data bank of frequencies set along the quarters of circles of a golden spiral developing from the point of origin and increasing with the logarithm of *phi* up to Planck frequency."

In effect, the ISS, or *Infinite Spiral Staircase*, a golden spiral, follows the Fibonacci sequence (Fn) from the Planck length (ℓP) toward the X-Point of origin of our universe, with the spiral's radii and wavelengths tending to the infinitely small, and frequencies tending to the infinitely high. And as we know, the velocity is function of the frequency, thus leading, in the spectrum of states (the radii of the quarter-circles of the ISS) reaching to the X-Point of origin, to velocities tending toward the infinite.

6. A Thought Experiment in Translocation of Information

Despite Richard Hoagland's old but still relevant complaint that "there seemed [to be] no testable, physical *proof* of 'hyperdimensional physics,'" here is a thought-experiment that, would it work and allow the transmission of information or of a matter system via the CSR-HD, would become a tangible and resonant proof of the reality of the HD. Of course the experimental design should be strongly shielded against any EM energy or possible carrier wave, especially the 2 gates of the system.

Here is the problem we aim at solving. How can we conceive and construct a machine that would allow us:

1. to transfer complex info from one place (in the 4D world) to another one too far to be reachable at the speed of light, by using the HD quasi-instant connection via sygons? (and eventually, in the long term, a network of these 'transfer gates').

2. to transfer matter systems (including organisms) instantly across space via the HD?

7.1. The principles of the Rhythm two-gate system

* Create 2 perfectly resonant gates. For the research phase, and to entangle the two gate-systems, these two gates G1 & G2 (shielded) would be set in proximity to each other, in the same shielded room. When positive results are registered, the two gates are set for testing at a distance, for example at the tip of two nearby hills, a few miles distant, with a flatten top. Their configuration should be exactly identical, and they need to be shielded from each other and from any EM outside perturbation.

* To instantiate the Center-Circle HD, the apparatus has to be set either (*a*) as a high velocity coherent beam rotating in a *torus*, of the highest possible frequency, or (*b*) as a coherent field, a Bose-Einstein condensate (BEC) standing wave on a *round surface*.

* To activate Rhythm HD, the high frequency coherent beam (in *a*) or the BEC's standing wave (in *b*) has to be modulated by either a complex and precise musical rhythmic melody (say, Steve Reich's 18 musicians), or a complex symmetrical design or exactly spatially-inverted symmetrical design (such as some chaos-based crop circles). And secondly, the rotation speed in the torus (angular momentum), the frequency and the melody in G1 have to be in perfect sync with that of G2.

* The "tune-in phase" of the 2 gates G1 & G2 (creating their HD entanglement) will always be the same for the 2 gates (once the optimal configuration has been found); it demands, in each perfectly identical gates, to start the system at a synchronized time for the launching of the coherent energy beam or field, and 'play' the modulated melody or design for a short but exact time.

* The "transfer phase" uses a second replay of the modulation in G1 & G2, to which is superposed, at the sending gate G1, the message as a melody, or as a superposed design.

* The entangled gate G2 (at the receiving end) should immediately be affected and record the superposed message, at the very time it is sent by the sender gate, whatever the distance between the gates. In other words, G2 should be able to detect and extract a distortion of its tune-in melody, which is precisely the message that's being sent.

7.3. Networks of Rhythm-HD gates

- Each entangled gate can function as the sending or receiving end in precise periods (the protocol is just inverted).

- A gate can be entangled with any number of other gates, two by two, each pair having its specific Tune-in melody/design acting as the connection-activating code. If a gate A was entangled with gate B and C…n, then all the entangled gates would receive the information simultaneously and instantly, but the precise sending of a matter- or bio-system (a translocation) would be impossible, as it would land at any one gate by pure chance alone.

- Networks of gates for translocation: Gate A would be entangled with gate B via Tune-in (Ti) melody AB, A is entangled with C via Ti-AC, etc. The selection of an AC transfer would trigger the Ti-AC.

- To translocate a system, large gates about 100 yards/meters in the form of a vertically set torus would be necessary. The system will be made to cross the torus and will emerge at the other location.

- To translocate a person, same configuration. It is hypothesized that the whole syg-field of the person (and that includes their bodies) will relocate at the receiving gate.

Conclusion

To conclude, I would like to stress that it is only because the hyperdimension in ISST ties together hyperspace, hypertime and hyperconsciousness that its virtual strings/particles, the sygons, may carry a complete information about such complex systems as a human being with a mind, a brain, and a body, or about a planet with its billion inhabitants. And similarly, the triune CSR-HD and all the sygons in our universe (HD and quantum-spacetime regions) are able to carry all the information about a universe in which thrive innumerable intelligent civilizations and complex beings.

The universe is gigantic, both at the macro and micro scales; in fact I'm starting to visualize that it extends as immensely below Planck scale as it does above it. We know that aliens are visiting us since ages, having left many a trace in 'impossible' monolithic architectures, and that our ancestors have recorded their existence on Earth with petroglyphs and cave paintings. While any sophisticated futurist propulsion or spaceflight system using a motion through space (such as anti-grav, the use of gravitational slingshot, solar sails…) could be part of their advanced technology, these would be mostly of a 'local' usage, and in no way would it allow interstellar communication or travel in any workable time-lapse.

It makes it all the more probable that they have discovered and mastered the fantastic dynamics and energy of the hyperdimension that can annull space and allow instant communication and translocation of beings or crafts at interstellar distances.

I'm deeply convinced that humanity's next scientific and philosophy paradigm is a holistic integration of cosmic and personal consciousness with physics and cosmology, and that this can and will be achieved solely via a hyperdimensional framework that includes consciousness at the origin and pervading the universe. This will be our gate to meeting other intelligent civilizations and interacting with them in a cosmic age of Earth.

Acknowledgments: This article was originally written for the 2019 MUFON project "Great Questions in Ufology" spearheaded by Dan Wright (editor).

Received February 22, 2020; Accepted March 30, 2020

References

Brandenburg, JE. *Beyond Einstein's Unified Field. Gravity and Electro-magnetism Redefined*. Kempton, Ill.: Adventures Unlimited Press, 2011.

———. *Life and Death on Mars: The New Mars Synthesis*. Kempton, Ill.: Adventures Unlimited Press, 2011.

Brandenburg, JE. & Hardy, CH. 2016. "Entropic Gravity in Pre-spacetime & the ISS Theory of a Cosmic Information Field." *Prespacetime Journal 7(5)*, 828-838. (9 April 2016). http://prespacetime.com/index.php/pst/article/view/968/944

Carr, B. 2010. "Seeking a New Paradigm of Matter, Mind and Spirit." *Network Review*, Spring & Summer.

———. *Universe or multiverse*. Cambridge, UK: Cambridge Univ. Press, 2009.

Do T., Hees A., Ghez A., et al. 2019. "Relativistic redshift of the star S0-2 orbiting the Galactic Center supermassive black hole." *Science,* Vol. 365, Issue 6454, pp. 664-668. 16 Aug 2019. DOI: 10.1126/science.aav8137

Fletcher L.N., et al. 2018. "A hexagon in Saturn's northern stratosphere surrounding the emerging summertime polar vortex." *Nature Communications* volume 9, Article number: 3564 (2018). *https://www.nature.com/articles/s41467-018-06017-3#Sec8*

Guth AH. 1997. *The Inflationary Universe*. Reading, Ms: Perseus Books.

Hardy, C.H. 2015b. "The quantum vacuum as a boundary to a hyperdimension: the ISST hypothesis." https://independent.academia.edu/ChrisHHardy/Papers

———. 2016. "ISS Theory: Cosmic Consciousness, Self, and Life Beyond Death in a Hyperdimensional Physics." *J. of Consciousness Exploration & Research* (JCER) Vol 7, No 11, pp. 1012-1035, Dec 2016. https://independent.academia.edu/ChrisHHardy/Papers

———. 2017. "Nonlocal consciousness in the universe: panpsychism, psi & mind over matter in a hyperdimensional physics." *Journal of Nonlocality 5(1),* June 2017. Special Issue. The other singularity: Psi and the Nonlocal Mind (edited by Ben Goertzel, Ph.D). https://independent.academia.edu/ChrisHHardy/Papers

———. *Cosmic DNA at the Origin: A Hyperdimension before the Big Bang. The Infinite Spiral Staircase Theory*. USA: CreateSpace, 2015.

———. *Space Allies*. (Exopolitics Sci-Fi series) USA: CreateSpace, 2017.

———. *The Sacred Network*. Rochester, Vt: Inner Traditions, 2011.

Hoagland, R. *Hyperdimensional Physics, Part 1:* http://www.enterprisemission.com/hyper1.html

Hu H. & Wu M. 2013. "What Is Quantum Gravity? What Is Graviton?" *Prespacetime Journal* 4 (11), pp. 1003-1026 (Dec 2013).

Jung CG. *Synchronicity: An acausal connecting principle*. The collected works of C.G. Jung: Vol. 8. (Bollingen Series, XX), Princeton, NJ: Princeton Univ. Press, 1960.

Jung CG, & Pauli W. *The Interpretation of Nature and the Psyche*. NY: Pantheon Books, 1955.

Kaku M. *Hyperspace: A Scientific Odyssey Through Parallel Universes, Time Warps, and the 10th Dimension*. New York: Anchor, 1994.

Magueijo, J. *Faster Than the Speed of Light: The Story of a Scientific Speculation*. Reading, Ms: Perseus Books, 2003.

Matson John. 2012. "Quantum teleportation achieved over record distances." *Nature News*, 13 August 2012. doi:10.1038/nature.2012.11163

Maxwell, James C. *A Dynamical Theory of the Electromagnetic Field*. (original publishing 1865)

———. *Treatise on Electricity and Magnetism*. 3rd Ed., 2 Vol. New York: Dover, 1954.

Pauli Wolfgang. *Exclusion principle and quantum mechanics*. Nobel Lecture, December 13, 1946. Full text on nobelprize.org. *pauli-Nobel%20lecture%20on%20Excl%20Princ%20(1946).pdf*

Randall Lisa. *Warped Passages: Unraveling the Mysteries of the Universe's Hidden Dimensions*. New York: HarperCollins, 2005.

Randall L, & Sundrum R. 1999. "An alternative to compactification." Physical Review Letters 83: 4690-93.

Riemann, Georg. "On the Hypotheses Which Lie at the Foundation of Geometry." Presentation made in 1854 at the University of Göttingen in Germany. Whole translation at https://books.google.co.in/books?id=BtZM33PzV9UC&pg=PA652&lpg=PA652&dq=Riemann,+Georg.+%22On+the+Hypotheses+Which+Lie+at+the+Foundation+of+Geometry

Sarfatti, J. *Super Cosmos; Through struggles to the stars. (Space-Time and Beyond III)*. Bloomington, In.: Author House, 2006.

Smolin, Lee. *The life of the cosmos*. New York: Oxford Univ. Press, 1997.

———. *The Trouble with Physics*. Boston, Ms: Houghton Mifflin Harcourt, 2006.

Vallee, J. *Fastwalker: A novel*. Berkeley, CA: North Atlantic Books,/Frog Ltd., 1996.

———. *Forbidden Science: Journals 1957-1969*. Berkeley, CA: North Atlantic Books, 1992.

———. *Passport to Magonia*. Washington, D.C.: H. Regnery Co., 1969.

Van Flandern, Tom. 1998. "The Speed of Gravity – What the Experiments Say." *Physics Letters A* 250:1-11. *http://www.ldolphin.org/vanFlandern/gravityspeed.html*

———. "Allais gravity and pendulum effects during solar eclipses explained." *Physical Review D* 67(2). Jan. 2003.

ISSN: 2153-8301   Prespacetime Journal   www.prespacetime.com
Published by QuantumDream, Inc.

Article

Linear Codes over the Family of Finite Rings A_t

Abdullah Dertli[*1] & Yasemin Cengellenmis[2]

[1]Dept. of Math., Faculty of Arts & Sci., Ondokuz Mayıs Univ., Samsun, Turkey
[2]Dept. of Math., Faculty of Science, Trakya University, Edirne, Turkey

Abstract

In this paper, we study the linear codes over the family of finite rings A_t. We define a Gray map Ω, which is an isometry map and linear. Furthermore, we study some properties of MDS codes over A_t and some examples.

Keywords: Linear codes, MDS codes, map.

1. Introduction

The most important family of finite rings for algebraic coding theory is linear codes. Recently, linear codes have raised interest since they have structure that makes to code and decode. MDS codes over the fields are one of the most important in coding theory and they are often the optimal codes for a particular of parameters. Some authors used different methods in order to obtain linear codes and MDS codes, [1-9].

In the present paper, the linear codes and MDS codes over the family of the rings A_t are studied, which is motivated by the previous works.

2. Preliminaries

A family of the finite rings $A_t = Z_4[u_1, u_2, ..., u_t]/\langle u_i^2 = u_i, u_i u_j = u_j u_i \rangle$, where $i, j = 1, ..., t$ contains the commutative rings with characteristic 4 and cardinality 4^{2^t}. The finite rings of the family are written as recursively

$$A_r = A_{r-1} + u_r A_{r-1}$$

where $r = 1, 2, ..., t$ and $A_1 = Z_4 + u_1 Z_4, u_1^2 = u_1$, where $A_0 = Z_4$.

We defined the Gray map recursively as follows

[*]Correspondence: Abdullah Dertli, Dept. of Math., Faculty of Arts & Sci., Ondokuz Mayıs Univ., Turkey.
E-mail: abdullah.dertli@gmail.com

$$\varphi_i : A_i \to A_{i-1}^2$$

$$\upsilon_{i-1} + u_i \rho_{i-1} \mapsto \left(\upsilon_{i-1}, \upsilon_{i-1} + \rho_{i-1} \right)$$

where $i = 2,...,t$ and

$$\varphi_1 : A_1 \to A_0^2$$

$$\upsilon_0 + u_1 \rho_0 \mapsto \left(\upsilon_0, \upsilon_0 + \rho_0 \right)$$

where $A_0 = Z_4$.

The Gray map φ_t can be extended to A_t^n, naturally.

Let $B \subseteq \{1,2,...,t\}$ and $u_B = \prod_{i \in B} u_i$. In particular $u_\varnothing = 1$. Each element of A_t is of the form $\sum_{B \in P_t} \vartheta_B u_B$, where $\vartheta_B \in Z_4$, P_t is the power set of the set $\{1,2,...,t\}$. For $M, N \subseteq \{1,2,...,t\}$ we have that $u_M u_N = u_{M \cup N}$ which gives that $\sum_{M \in P_t} \alpha_M u_M \cdot \sum_{C \in P_t} \beta_C u_C = \sum_{E \in P_t} \left(\sum_{M \cup C = E} \alpha_M \beta_C \right) u_E$.

Let

$$e_{u_\varnothing} = 1 + (-1)^{|B|} \sum_{B \in P_t} u_B$$

The number of $e_{u_\varnothing}$ is $\binom{t}{0}$.

$$e_{u_i} = u_i + (-1)^{|B|+1} \sum_{\substack{i \in B \in P_t \\ |B| \geq 2}} u_B$$

for $i = 1,2,...,t$. The number of e_{u_i} is $\binom{t}{1}$.

$$e_{u_i u_j} = u_i u_j + (-1)^{|B|+2} \sum_{\substack{i,j \in B \in P_t \\ |B| \geq 3}} u_B$$
$$\scriptstyle i < j$$

for $i,j = 1,2,...,t$. The number of $e_{u_i u_j}$ is $\binom{t}{2}$.

$$e_{u_i u_j u_s} = u_i u_j u_s + (-1)^{|B|+3} \sum_{\substack{i,j,s \in B \in P_t \\ |B| \geq 4}} u_B$$
$$\scriptstyle i < j < s$$

for $i,j,s = 1,2,...,t$. The number of $e_{u_i u_j u_s}$ is $\binom{t}{3}$.

$$\text{\Taurus}$$

$$e_{u_1 u_2 ... u_t} = u_1 u_2 ... u_t$$

ISSN: 2153-8301

Prespacetime Journal
Published by QuantumDream, Inc.

www.prespacetime.com

The number of $e_{u_1 u_2 \ldots u_t}$ is $\binom{t}{t}$.

Then we have, $\sum_{B \in P_t} e_{u_B} = 1, \left(e_{u_B} \right)^2 = e_{u_B}, e_{u_B} e_{u_A} = 0$ if $A \neq B$ for any $A, B \subseteq \{1, 2, \ldots, t\}$. Hence $A_t = \bigoplus_{B \in P_t} e_{u_B} A_t \cong \bigoplus_{B \in P_t} e_{u_B} Z_4$. So, every element z of A_t can be uniquely expressed as $z = \sum_{B \in P_t} a_{u_B} e_{u_B}$, where $a_{u_B} \in Z_4$.

Example 1: Let t be 3. Then
$$A_3 = Z_4 + u_1 Z_4 + u_2 Z_4 + u_3 Z_4 + u_1 u_2 Z_4 + u_1 u_3 Z_4 + u_2 u_3 Z_4 + u_1 u_2 u_3 Z_4$$
Consider the idempotent elements of A_3 below

$$e_{u_\varnothing} = e_1 = 1 - u_1 - u_2 - u_3 + u_1 u_2 + u_1 u_3 + u_2 u_3 - u_1 u_2 u_3$$

$$e_{u_1} = u_1 - u_1 u_2 - u_1 u_3 + u_1 u_2 u_3$$

$$e_{u_2} = u_2 - u_1 u_2 - u_2 u_3 + u_1 u_2 u_3$$

$$e_{u_3} = u_3 - u_1 u_3 - u_2 u_3 + u_1 u_2 u_3$$

$$e_{u_1 u_2} = u_1 u_2 - u_1 u_2 u_3$$

$$e_{u_1 u_3} = u_1 u_3 - u_1 u_2 u_3$$

$$e_{u_2 u_3} = u_2 u_3 - u_1 u_2 u_3$$

$$e_{u_1 u_2 u_3} = u_1 u_2 u_3$$

We can also define the Gray map as follows,
$$\Psi_t : A_t \to Z_4^{2^t}$$
$$z = \sum_{B \in P_t} a_{u_B} e_{u_B} \mapsto \Psi_t(z) = \Upsilon$$

where $\Upsilon = \left(\begin{array}{c} \sum_{B = \varnothing} a_{u_B}, \sum_{B \subseteq \{1\}} a_{u_B}, \sum_{B \subseteq \{2\}} a_{u_B}, \ldots, \sum_{B \subseteq \{t\}} a_{u_B}, \sum_{B \subseteq \{1,2\}} a_{u_B}, \sum_{B \subseteq \{1,3\}} a_{u_B}, \\ \ldots, \sum_{\substack{B \subseteq \{i,j\} \\ i < j}} a_{u_B}, \sum_{B \subseteq \{1,2,3\}} a_{u_B}, \ldots, \sum_{\substack{B \subseteq \{i,j,s\} \\ i < j < s}} a_{u_B}, \ldots, \sum_{B \subseteq \{1,2,\ldots,t\}} a_{u_B} \end{array} \right), \ a_{u_B} \in Z_4.$

The Gray map Ψ_t can be extended from A_t^n, naturally.

Example 2: Let $k = 3$. Then
$$\Psi_3 : A_3 \to Z_4^8$$
$$z = \sum_{B \in P_3} a_{u_B} e_{u_B} \mapsto \Psi_3(z) = \Upsilon$$

where
$$\Upsilon = \begin{pmatrix} a_1, a_1 + a_{u_1}, a_1 + a_{u_2}, a_1 + a_{u_3}, a_1 + a_{u_1} + a_{u_2} + a_{u_1 u_2}, \\ a_1 + a_{u_1} + a_{u_3} + a_{u_1 u_3}, a_1 + a_{u_2} + a_{u_3} + a_{u_2 u_3}, \\ a_1 + a_{u_1} + a_{u_2} + a_{u_3} + a_{u_1 u_2} + a_{u_1 u_3} + a_{u_2 u_3} + a_{u_1 u_2 u_3} \end{pmatrix}.$$

The Lee weight on Z_4, denoted w_L, is defined as

$$w_L(l) = \begin{cases} 0, & l = 0 \\ 1, & l = 1 \text{ or } 3 \\ 2, & l = 2 \end{cases}$$

For any $l = \sum_{B \in P_t} a_{u_B} e_{u_B} \in A_t$, the Lee weight of l is defined as

$$w_L(l) = w_L(\Psi_t(l)) = \sum_{i=1}^{2^t} w_L(l_i)$$

where $\Psi_t(l) = (l_1, l_2, ..., l_{2^t})$, $l_i \in Z_4, i = 1, 2, ..., 2^t$. The Lee weight of a vector $a = (a_1, ..., a_n) \in A_t^n$ is defined to be sum of the Lee weight of its components, that is $w_L(a) = \sum_{i=1}^{n} w_L(a_i)$. Moreover, for any $a_1, a_2 \in A_t^n$, the Lee distance between a_1 and a_2 is defined as $d_L(a_1, a_2) = w_L(a_1 - a_2)$.

Theorem 1: The Gray map Ψ_t is a linear and distance preserving map.

3. Methods

3.1 Linear Codes Over A_t

A non empty subset $\Im \subset A_t^n$ is called linear code over A_t if $\Im$ is a submodule of A_t.

Let $l = (l_0, l_1, ..., l_{n-1})$ and $\varsigma = (\varsigma_0, \varsigma_1, ..., \varsigma_{n-1})$ be two vectors in A_t^n. The Euclidean inner product of l and ς is defined

$$\langle l, \varsigma \rangle = \sum_{j=0}^{n-1} l_j \varsigma_j$$

where the operations are performed in the ring A_t.

Dual of the code $\Im \subseteq A_t^n$ is the code

$$\Im^{\perp} = \left\{ l \in A_t^n \ : \ \langle l, \varsigma \rangle = 0 \text{ for all } \varsigma \in \Im \right\}$$

If $\Im \subseteq \Im^{\perp}$, a code $\Im$ is said to be self orthogonal and if $\Im = \Im^{\perp}$, self dual.

Let

$$R_1 \oplus R_2 \oplus \ldots \oplus R_{2^t} = \left\{ r_1 + r_2 + \ldots + r_{2^t} \; : \; r_i \in R_i, i = 1, 2, \ldots, 2^t \right\}$$

$$R_1 \otimes R_2 \otimes \ldots \otimes R_{2^t} = \left\{ \left(r_1, r_2, \ldots, r_{2^t} \right) \; : \; r_i \in R_i, i = 1, 2, \ldots, 2^t \right\}$$

Define the codes C_{u_B}, as follows

$$C_{u_\varnothing} = C_1 = \{ a_{u_\varnothing} \in Z_4^n \; : \; \exists \; a_{u_B} \in Z_4^n, \sum_{B \neq \varnothing} a_{u_B} e_{u_B} \in \mathfrak{I} \}$$

$$C_{u_1} = \{ a_{u_1} \in Z_4^n \; : \; \exists \; a_{u_B} \in Z_4^n, \sum_{\substack{B \in P_t \\ B \neq \{1\}}} a_{u_B} e_{u_B} \in \mathfrak{I} \}$$

$$C_{u_2} = \{ a_{u_2} \in Z_4^n \; : \; \exists \; a_{u_B} \in Z_4^n, \sum_{\substack{B \in P_t \\ B \neq \{2\}}} a_{u_B} e_{u_B} \in \mathfrak{I} \}$$

$$\vdots$$

$$C_{u_t} = \{ a_{u_k} \in Z_4^n \; : \; \exists \; a_{u_B} \in Z_4^n, \sum_{\substack{B \in P_t \\ B \neq \{t\}}} a_{u_B} e_{u_B} \in \mathfrak{I} \}$$

$$C_{u_1 u_2} = \{ a_{u_1 u_2} \in Z_4^n \; : \; \exists \; a_{u_B} \in Z_4^n, \sum_{\substack{B \in P_t \\ B \neq \{1,2\}}} a_{u_B} e_{u_B} \in \mathfrak{I} \}$$

$$\vdots$$

$$C_{u_1 u_2 \ldots u_t} = \{ a_{u_1 u_2 \ldots u_t} \in Z_4^n \; : \; \exists \; a_{u_B} \in Z_4^n, \sum_{\substack{B \in P_t \\ B \neq \{1,2,\ldots,t\}}} a_{u_B} e_{u_B} \in \mathfrak{I} \}$$

The number of C_{u_B} is 2^t. Clearly, C_{u_B} is a linear code of length n over Z_4. $\mathfrak{I}$ can be uniquely decomposed into

$$\mathfrak{I} = \bigoplus_{B \in P_t} e_{u_B} C_{u_B}$$

and hence we have $\left| \mathfrak{I} \right| = \prod_{B \in P_t} \left| C_{u_B} \right|$.

Theorem 2: Let $\mathfrak{I} = \bigoplus_{B \in P_t} e_{u_B} C_{u_B}$ be a linear code of length n over A_t. Then the dual $\mathfrak{I}^\perp = \bigoplus_{B \in P_t} e_{u_B} C_{u_B}^\perp$ is also a linear code of length n over A_t.

Theorem 3: If $\mathfrak{I}$ is a (n, M, d_L) linear code over A_t, then $\Psi_t(\mathfrak{I})$ is a $(2^t n, M, d_L)$ linear code over Z_4.

Theorem 4: Let $\mathfrak{I}$ be a linear code of length n over A_t. Then $\Psi_t(\mathfrak{I}) = \bigotimes_{B \in P_t} C_{u_B}$.

Theorem 5: Let $\mathfrak{I}$ be a linear code of length n over A_t. If $\mathfrak{I}^\perp$ is dual of $\mathfrak{I}$, then $\Psi_t(\mathfrak{I})^\perp = \Psi_t(\mathfrak{I}^\perp)$. Moreover, if $\mathfrak{I}$ is self dual, so is $\Psi_t(\mathfrak{I})$ over Z_4.

3.2 Cyclic Codes Over A_t

In [7], the structure of cyclic codes of length n over Z_4 are determined. By using this, we will obtain the structure of cyclic codes over A_t.

Theorem 6: Let $\mathfrak{I} = \underset{B \in P_t}{\oplus} e_{u_B} C_{u_B}$ be a linear code over A_t. Then $\mathfrak{I}$ is a cyclic code over A_t if and only if C_{u_B} are cyclic codes over Z_4 for all $B \in P_t$. Moreover, if $\mathfrak{I}$ is a cyclic code over A_t, then

$$\mathfrak{I} = \left\langle e_1 f_1(x), e_{u_1} f_{u_1}(x), ..., e_{u_t} f_{u_t}(x), e_{u_1 u_2} f_{u_1 u_2}(x), ..., e_{u_1 u_2 ... u_t} f_{u_1 u_2 ... u_t}(x) \right\rangle$$

where $f_{u_B}(x)$ are generator polynomials of C_{u_B}, for all $B \in P_t$, respectively.

Let C_{u_B} be cyclic codes ove Z_4, n odd and $f_{u_B}(x) = g_{u_B}(x), 2a_{u_B}(x) = g_{u_B}(x) + 2a_{u_B}(x)$. Then the cardinality of C_{u_B} is $4^{n - \deg g_{u_B}(x)} 2^{\deg g_{u_B}(x) - \deg a_{u_B}(x)}$.

Corollary 1: Let $\Psi_t(\mathfrak{I}) = \underset{B \in P_t}{\otimes} C_{u_B}$ be a linear code of odd length over Z_4. Then the cardinality of $\Psi_t(\mathfrak{I})$ is $4^{\underset{B \in P_t}{\sum}(n - \deg g_{u_B}(x))} 2^{\underset{B \in P_t}{\sum}(\deg g_{u_B}(x) - \deg a_{u_B}(x))}$.

Let $\Psi_t(\mathfrak{I}) = \underset{B \in P_t}{\otimes} C_{u_B}$ be a linear code of odd length over Z_4 and d be the Lee distance of $\Psi_t(\mathfrak{I})$. Then $d = \min\{d_{u_B}, B \in P_t\}$.

3.3 MDS codes over A_t

In this section, we investigate some properties of MDS codes over A_t. $\mathfrak{I}$ is a linear code of length n over A_t, and d_{H_t} is the minimum distance, then

$$|\mathfrak{I}| \leq |A_t|^{n - d_{H_t} + 1}$$

So $d_{H_t} \leq n - \log_{|A_t|}|\mathfrak{I}| + 1$. This inequality is called Singleton bound (SB). If $\mathfrak{I}$, meet the Singleton bound, then $\mathfrak{I}$ are called MDS codes.

Lemma 1: Let $\mathfrak{I}$ be a linear code of length n over Z_4, the $\mathfrak{I}$ is a MDS code if and only if $\mathfrak{I}$ is either Z_4^n with parameters $(n, 4^n, 1)$ or $\langle 1 \rangle$ with parameters $(n, 4, n)$ or $\langle 1 \rangle^{\perp}$ with parameters $(n, 4^{n-1}, 2)$, where 1 denotes the all 1 vectors, [7].

We know that if $\mathfrak{I}$ is a linear code of length n over A_t, then

$$\mathfrak{I} = \underset{B \in P_t}{\oplus} e_{u_B} C_{u_B}$$

where C_{u_B} is a linear code of length n over Z_4.

Let d_{H_t} be the Hamming distance of $\Im$. Then $d_{H_t} = \min\{d_{H_{u_B}}\}$, where $d_{H_{u_B}}$ is Hamming distance of C_{u_B}. So the SB can be written as

$$d_{H_t} \leq n - \frac{1}{2^t} \sum_{B \in P_t} \log_{2^t} |C_{u_B}| + 1$$

Theorem 7: Let $\Im$ be a MDS codes over A_t,

   i.   If $d_{H_t} = 1$, then all of C_{u_B} are MDS codes with parameters $(n, 4^n, 1)$.

   ii.   If $d_{H_t} = 2$, then all of C_{u_B} are MDS codes with parameters $(n, 4^{n-1}, 2)$.

   iii.   If $d_{H_t} = n$, then all of C_{u_B} are MDS codes with parameters $(n, 4, n)$.

Proof (*i*) If $d_{H_t} = 1$, then $\sum_{B \in P_t} \log_{2^t} |C_{u_B}| = 2^t n$. Since $\Im$ is a MDS code over A_t, but $|C_{u_B}| \leq 4^n$, then the identity is true iff $|C_{u_B}| = 4^n$. Therefore $\Im$ is a $(n, 4^{2^t n}, 1)$ MDS code iff all of C_{u_B} are $(n, 4^n, 1)$ MDS codes.

The others are proved similar to (i).

Theorem 8: If $\Im$ is a MDS code over A_t, then there is at least one C_{u_B} be MDS code.

Proof It is proved that as in the proof the Theorem 4.3 in [1].

Corollary 2: $\Im$ is a MDS code over A_t iff all of C_{u_B} are MDS codes over Z_4 with same parameters.

Example 3: Let $n = 3, t = 6$.
$$x^3 - 1 = (x+3)(x^2 + x + 1) \in Z_4[x]$$
Let $C_{u_B} = \langle x^2 + x + 3 \rangle$ for $B \subseteq \{1,2,3,4,5,6\}$. We have $\Im = \bigoplus_{B \in P_6} e_{u_B} C_{u_B}$ and $\Psi_6(\Im)$ is a $\left(192, 4^{64} 2^{128}, 2\right)$.

Example 4: Let $n = 7, t = 8$.
$$x^7 - 1 = (x-1)(x^3 + 2x^2 + x + 3)(x^3 + 3x^2 + 2x + 3)$$
$$= g_1 g_2 g_3 \in Z_4[x]$$
Let $C_{u_B} = \langle g_2 g_3 + 2 \rangle$ for $B \subseteq \{1,2,3,4,5,6,7,8\}$. We have $\Im = \bigoplus_{B \in P_8} e_{u_B} C_{u_B}$ and $\Psi_8(\Im)$ is a

$$\left(1792, 4^{256}2^{1536}, 6\right).$$

Example 5: Let $n = 9$.

$$x^9 - 1 = \left(x+3\right)\left(x^2 + x + 1\right)\left(x^6 + x^3 + 1\right)$$
$$= g_1 g_2 g_3 \in Z_4[x]$$

Let $\quad C_{u_B} = \langle g_2 g_3 + 2g_2 \rangle \quad$ for $\quad B \subseteq \{1, 2, ..., t\}$. We have $\quad \Im = \bigoplus_{B \in P_t} e_{u_B} C_{u_B}$ and $\quad \Psi_t(\Im)$ is a

$$\left(9.2^t, 4^{2^t} 2^{3.2^{t+1}}, 3\right).$$

Received March 29, 2020; Accepted May 10, 2020

References

[1] Dertli, A., Cengellenmis, Y., "Some results on linear codes over the finite ring $Z_4 + uZ_4 + vZ_4$; MacWilliams identities, MDS codes", Journal of Science and Arts, 36: 209-215, (2016).

[2] Dougherty, S. T., Shiromoto, K., "MDS codes over Z_k", IEEE Trans. Inform. Theory, 46: 265-269, (2000).

[3] Dougherty, S. T., Shiromoto, K., "Maximum distance codes over rings of order 4", IEEE Trans. Inform. Theory, 47: 400-404, (2001).

[4] Gao, J., Gao, Y., "Some Results on Linear Codes over $Z_4 + vZ_4$", arXiv:1402.6 771v1, (2014).

[5] Guenda, K. and Gulliver, T.A., "MDS and self-dual codes over rings", Finite Fields Appl. 18(6): 1061-1075, (2012).

[6] Hammons, A. R., Kumar, V., Calderbank, A. R., Sloane, N. J. A., Sole, P., "The Z_4-linearity of Kerdock, Preparata, Goethals and related codes", IEEE Trans. Inf. Theory, 40: 301-319, (1994).

[7] Li, P., Guo, X., Zhu, S., "Some results of linear codes over the ring $Z_4 + uZ_4 + vZ_4 + uvZ_4$", arXiv:1601.04453v1, (2016).

[8] Shiromoto, K., "Singleton bounds for codes over finite rings", Journal of Algebraic Combinatorics, 12: 95-99, (2000).

[9] Wan, Z.-X., Quaternary codes, World Scientific Publishing, (1997).

Review Article

Detecting Gravitons on Black Hole Coalesence

Lawrence B. Crowell [1]

Abstract

This paper is a short review of an article on how quantum hair on black holes may produce measurable signatures in gravitational radiation. This may be a significant step in the understanding of quantum gravitation and some experimental support for theories thereof.

Keywords: Black Hole, coalesence, graviton, detection, quantum hair, gravitational radiation.

1   Introduction

It is increasingly common to hear that quantum gravitation, in particular superstring theory, has gone off into mathematical abstractions. While this is an issue with the complexity of theory, the real problem is one of scale. To produce an exciton of the quantum gravitational field, such as a quantum of black hole, requires transverse momenta or energy 16 orders of magnitude larger than the LHC. To produce a graviton, which is very weakly interacting, that can be at all detectable would require comparable energy. It is not hard to calculate that an LHC type of machine would have to encompass the Milky Way galaxy to achieve Planck energy scale interactions. It is then apparent there will be no Planck energy particle physics experiments.

However, nature may provide events and processes of sufficient power. The coalescence of black holes produces an enormous amount of gravitational radiation. Gravitational waves should be ultimately quantum mechanical, and the production of gravitational waves should have signatures of quantum mechanics[1]. Gravitational radiation is produced when the mass of two black holes is converted into gravitational radiation that carries off information. The theoretical limit for gravitational wave production is where the area of the two equal mass black holes equals the area of the merged black hole. This means $S_{tot} = 8m^2 = 4M^2$. The mass of the merged black hole is then $\sqrt{2}m$ and $(2 - \sqrt{2})m$ of mass energy is produced as gravitational radiation. This $\frac{\sqrt{2}}{2}\%$ of mass energy is a theoretical lower limit for no entropy change. The upper limit is for a final mass $M = 2m$ and maximal entropy increase, where no gravitational radiation is produced. The coalescence of extremal black holes would have this result. LIGO measurements indicate a usual 5% of total mass released as gravitational radiation. This massive amount of gravitational radiation is ultimately quantum mechanical and composed of gravitons. Signatures of these gravitons should then exist in this radiation.

Entropy increase is a measure of hair on the event horizon. For extremal black holes with $r = a$ or Q the horizon has maximal hair and gravitational radiation is not produced. However, even for Schwarzschild black holes of equal mass that have no hair in the form of charge or angular momentum a coalescence results in a black hole with a mass greater than $\sqrt{2}m$. Entropy will in general increase. This is ultimately due to the fact the stretched horizon of a black hole has quantum hair. This is holographic quantum information of what composes the black hole. This quantum hair will then result in a signature in gravitational radiation as gravitational memory[2].

This signature in gravitational waves results from an event horizon induced from of the Casimir effect. In the last 10^{-24} seconds before the horizons actually merge holographic information on the stretched horizon interact and contribute to the entropy of the coalescing black holes. This might appear to be a

---

[1]Correspondence: Lawrence B. Crowell, PhD, Alpha Institute of Advanced Study, 2980 FM 728 Jefferson, TX 75657 and 11 Rutafa Street, H-1165 Budapest, Hungary. Tel.: 1-903-601-2818 Email: lcrowell@swcp.com.

ISSN: 2153-8301   Prespacetime Journal   www.prespacetime.com

Published by QuantumDream, Inc.

tiny UV physics, but with the tortoise coordinate stretching of wavelengths $r^* = r - ln|r - 2m|$ for the radius $r = 2m + \lambda$ this quantum hair can be expanded into the .1 to 10^2 Hz range[1]. This is within the detector sensitivity of LIGO at the low end and the future eLISA space-based interferometer. This information will imprint itself as gravitational memory. This is where test masses are not restored to their initial configurations after the passage of a gravitational wave.

The production of gravitons in this manner is related to Hawking radiation as a form of exciton generation. Hawking radiation could be thought of as the production of a quantum black hole that pinches off the horizon of a black hole. By the time this quantum makes its way out of the gravitational well it is IR long wavelength radiation. This is a related process, but where the coalescence of black holes induces the rapid production of gravitons through the excitation of quantum hair on horizons. However, this is an explicit generation of gravitons and more closely related to quantum gravitation.

2 Quantum mechanics in a spacetime sandwich between horizons

The Reisner-Nordstrom metric for a charged black hole has a near horizon condition that is equivalent to $AdS_2 \times \mathbb{S}^2$[1]

$$ds^2 = \left(\frac{r}{m}\right)^2 dt^2 - \left(\frac{m}{r}\right)^2 dr^2 - m^2 d\Omega^2.$$

This is also the condition that Carroll, Johnson and Randall found for the extremal spacelike region in the Kerr metric[3]. This suggests a connection to anti-de Sitter geometry, and this is argued to be more completely the case for the sandwiched region between two black holes within 10^{-25} seconds of merging. This however must be argued without reference to an exact metric or solution.

The sandwich region is argued to be locally AdS_4 from two points[1]. The first is to look at two black holes that are far from a merger and approximately calculate the curvatures between them. The second argument is to look at the above metric but where the near horizon condition expands the area of the region with $m^2 \rightarrow r^2 + \rho^2$. This can be seen physically reasonable as the spherical nature of the metric is clearly perturbed and expanded. Also, the first approach indicates clearly there is a spatially hyperbolicity. It is then evident this sandwiched region is a deformed variant of AdS_4. To model this deformation the spatial hyperbolic space $H^3 \subset AdS_4$ in this region is mapped into a strip. The same could be seen geometrically if the antipodal points of a Poincar disk are sent to infinity. This is diagrammatically represented below

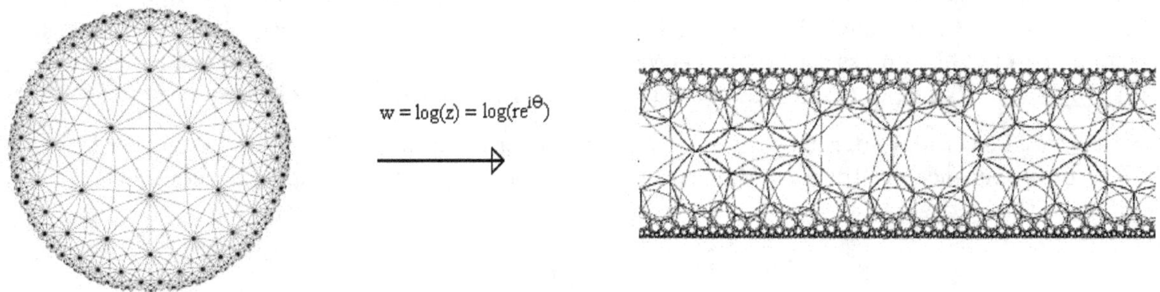

This map sends the Poincar disk to a strip, that in three dimensions is the hyperbolic sandwich. In a classical setting a particle trajectory is given by these arcs. This hyperbolic path is also the type of

trajectory a particle from Unruh radiation follows. This spatial sandwich is then similar to a Rindler wedge and the quantum hair on the event horizon generates quanta in this region. The most elementary quantum equations to analyze are the Dirac and Klein-Gordon equations. The analysis performed reduced the dimension by one to consider AdS_3 in 2 plus 1 spacetime. The near horizon condition for a near extremal black hole in 4 dimensions is considered for the BTZ black hole. This AdS_3 spacetime is then a foliations f hyperbolic spatial surfaces H^2 in time. These surfaces under conformal mapping are a Poincare disk. The motion of a particle on this disk are arcs that reach the conformal boundary as $t \rightarrow \infty$. This is then the spatial region we consider the dynamics of a quantum particle. This particle we start out treating as a Dirac particle, but the spinor field we then largely ignore by taking the square of the Dirac equation to get a Klein-Gordon wave[1].

Define the z and $\bar{z}$ of the Poincare disk with the metric

$$ds^2_{p-disk} \;=\; R^2 g_{z\bar{z}} dz d\bar{z} \;=\; R^2 \frac{dz d\bar{z}}{1 - z\bar{z}}$$

with constant negative Gaussian curvature $\mathcal{R} \;=\; -4/R^2$. This metric $g_{z\bar{x}} \;=\; R^2/(1 - \bar{z}z)$ is invariant under the $SL(2, \mathbb{R}) \sim SU(1, 1)$ group action, which, for $g \in SU(1, 1)$, takes the form

$$z \;\rightarrow\; gz \;=\; \frac{az + b}{\bar{b}z + \bar{a}}, \; g \;=\; \left(\begin{array}{cc} a & b \\ \bar{b} & \bar{a} \end{array} \right). \tag{2.1}$$

The Dirac equation $i\gamma^\mu D_\mu \psi + m\psi \;=\; 0$, for $D_\mu \;=\; \partial_\mu + iA_\mu$ on the Poincare disk has the Hamiltonian matrix

$$\mathcal{H} \;=\; \left(\begin{array}{cc} m & H_w \\ H_w^* & -m \end{array} \right) \tag{2.2}$$

for the Weyl Hamiltonians

$$H_w \;=\; \frac{1}{\sqrt{g_{z\bar{z}}}} \alpha_z \left(2D_z \;+\; \frac{1}{2}\partial_z(ln\ g_{z\bar{z}}) \right),$$

$$H_w^* \;=\; \frac{1}{\sqrt{g_{z\bar{z}}}} \alpha_{\bar{z}} \left(2D_{\bar{z}} \;+\; \frac{1}{2}\partial_{\bar{z}}(ln\ g_{z\bar{z}}) \right),$$

with $D_z \;=\; \partial_z + iA_z$ and $D_{\bar{z}} \;=\; \partial_{\bar{z}} + iA_{\bar{z}}$. here α_z and $\bar{\alpha}_z$ are the 2×2 Weyl matrices[1]. These wave equations are then a quantum version of the hyperbolic dynamics of Mirzakhani and Eskin[4].

With these two field sources and some approximations, such as considering the fields in the middle of the sandwich, the wave solution is given by a parabolic cylinder function plus a Laguerre polynomial. A numerical output of these is seen below. The parabolic cylinder function has as special cases harmonic oscillator solutions. This connects these analyses with the normal mode theory of Corda. This means that black hole quantum mechanics, at least to some fair degree of approximation, as dynamics that is similar to an atom. The emission and absorption of quanta by black holes shares features similar to an atom[5]. This follows similar analysis for normal or harmonic oscillator modes with black holes[6].The emission of quanta by this process is then a superposition of a hydrogen atom-like system and a harmonic oscillator. Black hole dynamics then connects with aspects of standard quantum physics.

The differential equations are conformal and solutions conformal invariant This is consistent with the AdS/CFT correspondence identified by Maldacena[7]. Further the parabolic cylinder function defines a path integral with the Lagrangian

$$\mathcal{L} \;\rightarrow\; \frac{1}{2}\partial_\mu \chi \partial^\mu \chi \;+\; \alpha \;+\; \frac{1}{2}\mu^2 \chi^2 \;+\; \frac{1}{4}\lambda \chi^4,$$

where $\frac{2}{3}\alpha \;\rightarrow\; \frac{1}{4}\lambda$. The functional derivatives are then

$$\left((p^2 + m^2)\frac{\delta}{\delta J} \;+\; \lambda\frac{\delta^3}{\delta J^3} \right) Z \;=\; -i\left\langle \frac{\delta S}{\delta \chi} \right\rangle,$$

ISSN: 2153-8301   Prespacetime Journal   www.prespacetime.com

Published by QuantumDream, Inc.

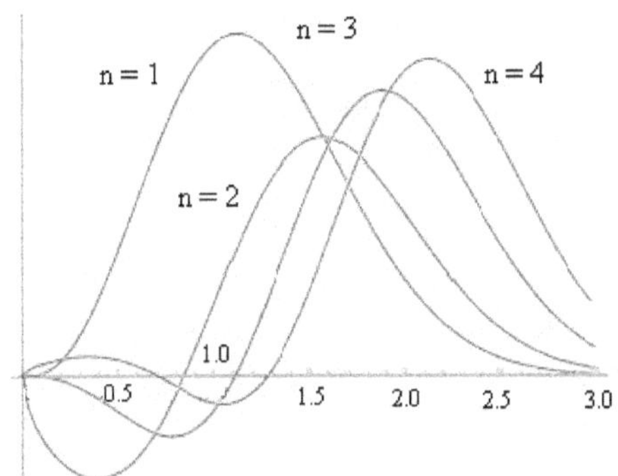

Parabolic cylinder function for n = 1, 2, ... 4 represented as a Hermite polynomial solution of the form $x^{1/4} e^{-x^2} H_n(x^2)$.

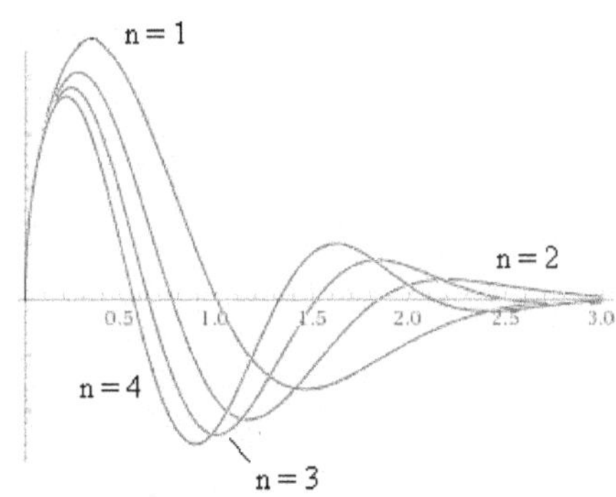

Laguerre wave function $x^{1/4} e^{-x^2} L_n^0(x^2)$ for hydrogen atomic-like states for n = 1, 2, 3, 4.

These are the wave function components contributed by the parabolic cylinder functions, or Hermite polynomials and the Laguerre polynomials. These depend on $x^2 = k\xi^2$ so the wave function is radial. These are not normalized.

This cubic form has three parabolic cylinder solutions. We may think of this as $ap + bp^3 = J$ and is a cubic equation for the source J that is annulled at three points. The correspond to distinct solutions with distinct paths. These three solutions correspond to three contours and define three distinct vacua. The overall action is a quartic function, which will have three distinct vacua, where one of these is the low energy physical vacua. It is worth noting this transformation of the problem has converted it into a system similar to the Higgs field[1].

The primary interest in these analyses is to explore how quantum mechanics of gravitation and in particular gravitons can be detected or inferred in the detection of gravitational radiation. These signatures will be found in BMS translations[2]. These could be detected by the eLISA program. This is a triangle of spacecraft in an orbit chasing Earth. Each spacecraft has a different inclination so the orbital planes of the three spacecraft is inclined relative to the ecliptic by about 0.33 degree. This defines a plane of the triangular spacecraft formation tilted 60 degrees from the plane of the ecliptic. Each spacecraft houses a laser interferometer and keeps it from being accelerated by external forces. The LISA Pathfinder mission to test this result was a success[8].

Quantum hair will generate information that reaches $\mathcal{I}^\infty$ as translations. This is BMS information that will adjust the metric of spacetime and extension the spatial relationship between the three interferometers.

The Bondi metric is [9]

$$ds^2 = du^2 - 2dudr + 2r^2\gamma_{z\bar{z}} + \frac{2m}{r}du^2 + rC_{zz}dz^2 + rC_{\bar{z}\bar{z}}d\bar{z}^2 + D^z C_{zz}dudz + D^{\bar{z}}C_{\bar{z}\bar{z}}dud\bar{z} + \ldots,$$

for $\gamma_{z\bar{z}} = 2(1 + z\bar{z})^{-2}$ the metric on the unit $\mathbb{S}^2$ sphere. The term m_B is the Bondi mass term, say for a BH and source of mass-energy propagating out to $\mathcal{I}^+$, where the coordinate u is defined. The terms C_{zz} and $C_{\bar{z}\bar{z}}$ determine Weyl curvature terms for gravitational wave propagating out. An Einstein field equation is

$$D_{\bar{z}}^2 C_{zz} - D_z^2 C_{\bar{z}\bar{z}} = 0,$$

which gives the simple solution $C_{zz} = D_z^2 C(z, \bar{z})$. Here $C(z, \bar{z})$ is a scalar potential. The change in

ISSN: 2153-8301   Prespacetime Journal   www.prespacetime.com
Published by QuantumDream, Inc.

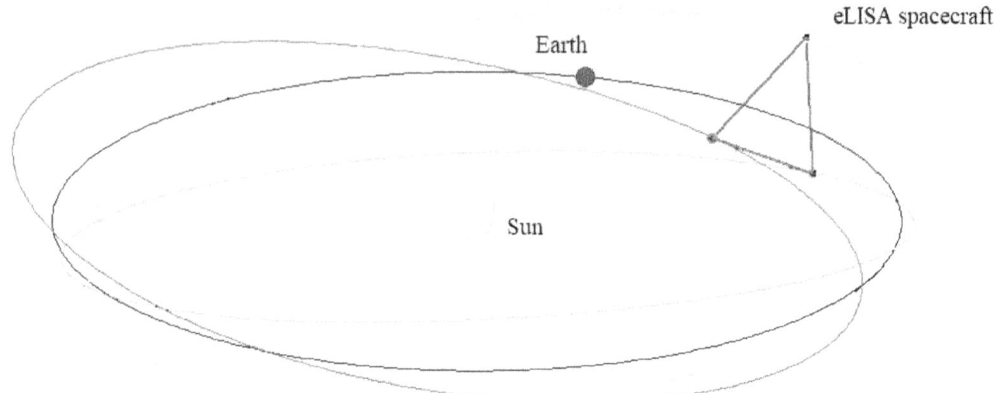

this potential is a change in Weyl curvature with the passage of a gravitational wave[9]. These define the information generated by quantum hair.

3   Discussion

This development clearly needs refinement. There is a need for explicit calculation of the BMS translations expected. The calculation was also performed with a high degree of approximation. There are of course no exact solutions possible, and further work may require numerical analysis. This may lead to future detection of signatures for gravitons. This with future experiments with quantization on the large and superposed metrics may yield some future experimental understanding of quantum gravitation.

The formalism with the near horizon anti-de Sitter space and conformal symmetry is in line with AdS/CFT correspondence and M-theory. This also leads with the connection to normal modes some general transformation of spacetime solutions via quantum modes. The AdS and dS spacetimes are separated by a light cone in one dimension larger. The de Sitter spacetime on the outside connects with the AdS inside the cone at $\mathcal{I}^\infty$. Thus ultimately the two share the same quantum information. Thus, while string theory has difficulties in de Sitter spacetime with a positive vacuum energy, the field theoretic content of AdS or CFT on the boundary connect to the quantum field theoretic in dS. Further, this all points to methods for understanding how different physical systems in general relativity transform into each other.

Received March 16, 2020; Accepted April 8, 2020

References

[1] C. Corda, L. Crowell, Quantum Hair on Black Holes, Entropy 2020, 22, 301; doi:10.3390/e22030301 https://www.mdpi.com/1099-4300/22/3/301

[2] M. Favata, "The gravitational-wave memory effect," Class. Quant. Grav. **27** 084036 (2010). https://arxiv.org/abs/1003.3486

[3] S. M. Carroll, M. C. Johnson, L. Randall, "Extremal limits and black hole entropy," JHEP 0911:109,(2009). https://arxiv.org/abs/0901.0931

[4] A. Eskin, M. Mirzakhani, "Counting closed geodesics in Moduli space," https://arxiv.org/abs/0811.2362

[5] C. Corda, Class. Quantum Grav. **32**, 195007 (2015).

[6] M. K. Parikh and F. Wilczek, Phys. Rev. Lett. **85**, 5042 (2000).

[7] J. M. Maldacena, "The Large N Limit of Superconformal Field Theories and Supergravity," Adv. Theor. Math .Phys. **2**, 231-252, (1998).

[8] www.lisamission.org

[9] A. Strominger and A. Zhiboedov, "Gravitational Memory, BMS Supertranslations and Soft Theorems," http://arxiv.org/abs/1411.5745

Essay

Painting, Baking & Non-Associative Algebra

Jonathan J. Dickau[*]

Abstract

There are analogies of the complications encountered when using the Octonions or other non-associative algebras that occur in everyday tasks and extend to the syntax of human language. The forced ordering and sequence of operations imposed by higher-order algebras is seen to be equivalent to its representations in these forms. Tasks like painting and baking involve cycles of action in a certain direction, undertaken in stages, where a new stage of a process can be undertaken only once certain steps are completed. I discuss briefly how this is exactly like octonion multiplication. This has relevance in topics from Cosmology to Computing and beyond.

Keywords: Octonion algebra, non-associativity, ordering of elements, sequence of operations.

Introduction

A defining feature of non-associative algebras is their strong directionality, involving both the forced ordering of elements and strict sequencing of operations. This is in contrast with familiar Maths where elements commute and can be associated arbitrarily. In arithmetic using ordinary or real numbers, both the associative and commutative rules can be applied to simplify calculations. However, if one needs to do both addition and multiplication in the octonions (with 8 dimensions) or other higher-dimensional algebras, the ordering of elements and the sequence of operations must be strictly observed to obtain the correct result. This is because the octonions are both non-commutative and non-associative.

What this means is that cycles of action flow in a particular direction and must be completed in a specific sequential order. This brings to mind everyday tasks like painting or baking, where directional cycles (with ordering of elements) allow one to complete a stage in a process, and then the stages are completed in a specific sequence to bring a process to fruition. It is not possible to put on the icing, bake the cake afterward, mix the ingredients, find and measure the ingredients after that, and then make the icing. This is because the ingredients must be blended before baking, the cake must be baked and have time to cool before icing it, and we want the icing on the outside with any filling on the inside. Similarly, placement of parentheses (or other grouping) must reflect and preserve the sense of what we want to end up on the outside or the inside, in non-associative Maths, just as the cake, filling, and icing ingredients are kept separate, and then used in a different part of the process.

---

[*] Correspondence: Jonathan J. Dickau, Independent Researcher. E-mail: jonathan@jonathandickau.com

Everyday Tasks and Cosmic Evolution

When we decided it was time to paint a few rooms in our house, I began the process by assessing what furniture had to be pulled away from the walls, and where it could go, which spots needed repair or preparation before we began, and so on. Sometimes I needed to remove things from shelves first, and to find storage places for those items, before I could move certain pieces of furniture safely. But it became clear early on that I needed to think from the end of the process, and then trace things back to what has to happen, in what order, to begin and then finish that process to completion. Final completion is hard sometimes. There are too many ways to end up in a *cul de sac* instead, and one must think several steps ahead to avoid literally painting oneself into a corner. But there are no shortcuts to obtaining the desired result, in many cases.

An extender pole or corner brush might allow one to get to hard-to-reach places, but there is a limit to how much extra reach it will provide. Nor can one decide after the space around it is painted that a cabinet needs to be moved, especially once it is blocked in by other furniture that was already moved. It is the same with non-commutative and non-associative algebras like the octonions. There is extreme optiony for octonion multiplication at the outset, with 480 different multiplication tables possible [1]. But once a starting place and direction is chosen, this reduces to one of 16 unique variations (8 left and 8 right-handed) [2] where the order and sequence is determined to the last detail, and where only one table must be applied throughout – until the last calculation.

The fact the octonions are neither commutative nor associative is bothersome for many who do Math and this has limited their study and usage. However, they may be essential for understanding the origin of the cosmos, discerning what happens at the rim of a black hole, or solving the problem of quantum gravity. Michael Atiyah said that a true theory of everything will almost certainly involve the octonions [3] and it may be hard to find "because we know the octonions are hard, but when you've found it, it should be a beautiful theory, and it should be unique." P.C. Kainen thinks the octonions being neither commutative nor associative is a blessing, not a curse, since it gives them unique evolutive properties useful to Physics [4].

Ordering of elements and sequencing of stages are essential for some physical processes, though physical law makes many processes reversible – because the same law works forward or backward in time. As more interactions occur, it becomes harder and harder to reverse any event or process. Once a new process stage has been entered, it is difficult or impossible to go back to what was before, or to choose the 'road not taken,' but instead further steps tend to accumulate into the next stage. This kind of ratcheting effect may explain the arrow of time. The onset of associativity in the early universe likely came with the emergence of fermionic particles during baryogenesis, because they are self-contained forms with a well-defined surface and thus an interior and exterior – so this pushed the cosmic creation process inexorably forward [5].

However, this property of processes being ratcheted forward once a stage or cycle is completed extends far beyond the origin of the cosmos, and it reaches into every area of life and human endeavor. So it is not surprising that this sense of process direction extends into human language, and into areas of thought far removed from Mathematics. Understanding that one needs to undertake directional cycles of action, and then move from one stage to another during a process (with a different character of action at each stage), allows for the planning and execution of complex tasks.

To model such processes mathematically, and thus optimize for maximum gain or progress, requires higher-dimensional thinking. But this enables planning that brings a process to its final completion, rather than being halted before then because some complication was not foreseen. The most important thing, in some cases, is care and precision. To take apart, clean, and reassemble an old-style clock, one must take great care indeed because things need to come apart in a specific order and sequence, and then go back together with the same steps exactly reversed, without having any pieces left over when you are done. And there are many examples in everyday life which require a similar sequential attention to detail.

The point is that we find the property of non-associativity to be more ubiquitous than one would expect, based upon the preponderance of Maths in Physics that use algebras that are both commutative and associative. And I stress here that higher-order algebras like the quaternions and octonions are not more disordered or random for losing the properties of commutativity and associativity, but rather there is a greater sense in both cases of following a specific order and sequence instead of choosing arbitrarily. Alain Connes first glimpsed that non-commutative algebra and geometry offer unexpected and amazing features not present in the static commutative case, back in the early 1970's. He summed this up in 2000, saying "noncommutative measure spaces evolve with time!" [6] and explaining there is a one parameter group of automorphisms driving this evolution.

This dynamism is present in the quaternions, but it is amplified and extended in the octonions and other non-associative algebras, so it becomes sequential evolution. Processes in the real world often progress through sequential stages of evolution, and it appears to be how the cosmos emerged as well, so this affirms there is a place for non-associative Maths in solving real world problems and in helping us understand Cosmology. It would appear that nature already makes broad use of them in every facet of our existence, so it is time we did too.

ISSN: 2153-8301   Prespacetime Journal   www.prespacetime.com
                                   Published by QuantumDream, Inc.

Conclusions

I have emphasized here that the utility of non-associative Maths recommends their broader use, even though this brings complications. The fact people avoid using them is quizzical, given what we have come to know about their usage. Computers make it easy to systematize the process to an extent allowing their extensive use without having to keep track of the sequential evaluation of terms. In other words, the computer lets us make use of a powerful new engine, without needing to build an engine ourselves every time we want to drive somewhere. Of course, it needs to be built from the inside out, piece by piece, in order to function as a smoothly-running machine.

However, people go to great lengths to avoid the complications introduced by non-associative Maths, even though it is the very same properties that make them complicated, which make them useful or interesting. John Baez wrote [7] that while the real numbers are seen as the dependable and respectable breadwinner and the complex numbers as a flashier but still respectable youngster, the quaternions are shunned because they are non-commutative, and the octonions are seen as the crazy old uncle nobody wants to talk about and would rather leave locked in the attic. I'd instead think of them as a wealthy and eccentric uncle nobody quite understands:

$$\mathbb{O} \supset \mathbb{H} \supset \mathbb{C} \supset \mathbb{R}$$

In terms of generality, the octonions are the granddaddy of all the normed division algebras. They are the most general of all the number types with those properties (division algebras with a norm). If we see the seven imaginaries of the octonions as axes of rotation, fixing four of seven axes yields the quaternions, fixing two of the remaining three yields the complex numbers, and the reals are obtained by fixing the remaining axis – reducing the variations in all directions orthogonal to the reals to zero. So we see that, the reals $\mathbb{R}$ are a subset of the complex numbers $\mathbb{C}$, which are a subset of the quaternions $\mathbb{H}$, and the quaternions are a subset of the octonions $\mathbb{O}$. This means numerical quantities we are most familiar with, which just sit there unchanging, are the product of quantities that contain more variation than sameness.

In fact, the octonions have seven dimensions of variation and one real extent. So this is hard for people to visualize, who have been told that numbers are the most predictable and dependable things that exist. We have done a disservice by educating our young people that *all* numerical quantities are static. My brief conversation with Tevian Dray at GR21 revealed that the appearance of non-commutative and non-associative terms is unavoidable in many regimes important to Gravity and Cosmology [8], which is like an elephant in the room [9] with researchers in that arena because even those few who know it is a problem don't know how to deal with it. We all need to learn how to work with numbers the way we paint and bake.

(Published in Prespacetime Journal| May 2020 | Volume 11 | Issue 3 | pp. 264-268)
Dickau, J. J., *Painting, Baking & Non-Associative Algebra*

Received April 23, 2020; Accepted May 23, 2020

References

1 Smith, F.D. 'Tony', Octonion Products and Lattices, where do the 480 octonion multiplication products come from?, http://www.tony5m17h.net/480op.html

2 Lockyer, Richard, 16 Octonion algebras, 480 representations, from 'Octonion Algebra – A presentation of the algebra and its connection to physics,' found at: http://octospace.com/files/16_Octonion_Algebras_-_480_Representations.pdf

3 Atiyah, Michael, From Quantum Physics to Number Theory, lecture at IMPA (2010) view at: https://www.youtube.com/watch?v=zCCxOE44M_M read more at: Wolchover, Natalie, The Peculiar Math That Could Underlie the Laws of Nature, *Quanta magazine* (July 2018), https://www.quantamagazine.org/the-octonion-math-that-could-underpin-physics-20180720/

4 Kainen, Paul C. ., An Octonion Model for Physics, proceedings of the 4th Conference on Emergence, Coherence, Hierarchy, and Organization (ECHO 4), Odense, Denmark, 2000, full-text at: https://faculty.georgetown.edu/kainen/octophys.pdf

5 Dickau, Jonathan J., The Origin of Time in Mandelbrot Cosmology, *Prespacetime Jour.*, 10, 3, (May 2019) pp. 348-359, https://prespacetime.com/index.php/pst/article/view/1568

6 Connes, Alain, Noncommutative Geometry Year 2000, pp. 481-559 in Alon N., Bourgain J., Connes A., Gromov M., Milman V. (eds) "Visions in Mathematics." Modern Birkhäuser Classics, Birkhäuser Basel, arXiv:math/0011193

7 Baez, John C. ., The Octonions, *Bull. Amer. Math. Soc.* **39** (2002), pp. 145-205, full text here: https://www.ams.org/journals/bull/2002-39-02/S0273-0979-01-00934-X/S0273-0979-01-00934-X.pdf, arXiv:math/0105155

8 Dray, Tevian, conversation with the author at GR21 in NYC, (July 2016)

9 Dickau, Jonathan J., Putting the Elephants to Work, FQXi contest essay (2017), available here: https://fqxi.org/community/forum/topic/2762

Essay

Is Gravity Curvature of Space-time?

Emre Dil[*]

Physics/Energy Engineering Dept., Beykent University, İstanbul, Turkey

Abstract

In this essay, we propose a semi-classical approach to support the interaction force nature of gravity as opposed to its curved geometrical nature. By giving the 3-D and 1-D mass configurations in corresponding space-times claimed as curved, we present there still requires a force concept in order to explain the gravity and to start the motion due to gravity. Consequently, we infer that the curvature interpretation of gravity does not describe the underlying natural phenomenon, but behaves as a mathematical tool to predict the natural phenomena correctly.

Keywords: Gravity, curvature, general relativity.

Is gravity really curvature of space-time or is it just able to be explained by the curvature? This deep question really requires a dare to ask. In a previous work it has been tried to present a support for that the curvature is not a fact of the gravity, but just a mathematical tool and model to explain it on a gravitationally deflected light due to a black hole (Dil 2019). There has been proposed a mass induction process in order to explain the gravitational deflection of light due to a massive object instead of the followed curved path by the light. In the literature some other attempts can also be found to support gravity as an interaction force rather than the space curvature concept (Friedman, 2016; Friedman & Steiner, 2016; Friedman & Steiner, 2017).

In this paper, we give another support to present the gravity is just an interaction force instead of the misunderstood space-time curvature concept. In the commonly accepted model, we assume a flat sheet of space on which a massive source object M is to be placed as in the Figure 1a. Because the massive object forms a curvature on the sheet representing the space, we assume that if another test object with mass m gets closer to the horizon it is attracted toward the source object. However, if we assume that the figure is looked from another perspective as in Figure 1b because there is no a certain above-below direction in space, it does not make any sense that the test object has to be attracted toward the left side. The sense of attraction of test object in Figure 1a is just an unfortunate misconception, since we intuitively expect the below side is the direction of gravity. In Figure 1b this misconception really vanishes.

---

[*]Correspondence: Emre Dil, Beykent University, İstanbul, Turkey. E-mail: emredil@beykent.edu.tr

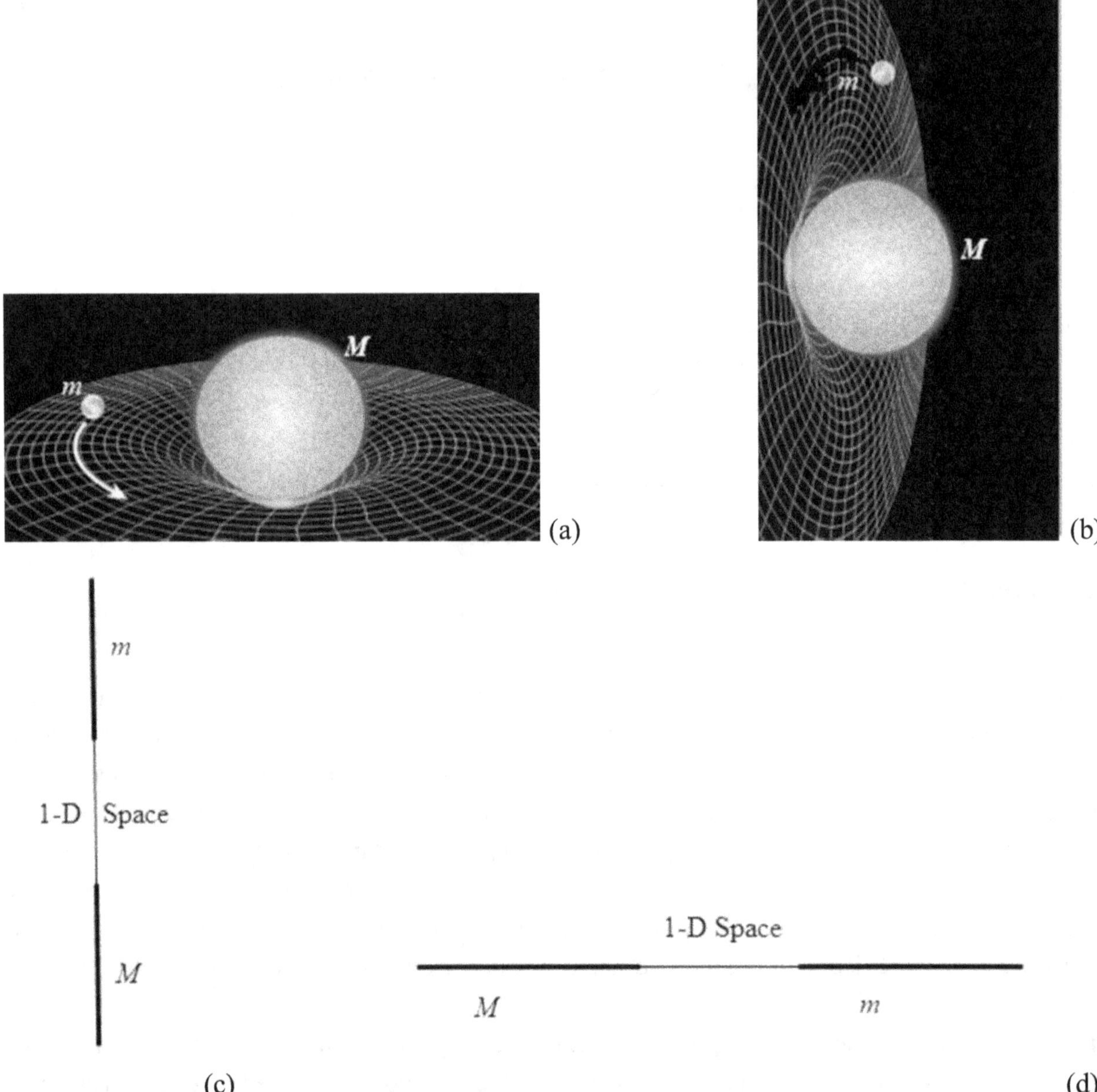

Figure 1. Source mass M and test mass m are placed into: (a) Space viewed from top, (b) Space viewed from right, (c) 1-D space viewed from front, (d) 1-D space viewed from side with strain curvatures on it.

The curvature cannot solely start the motion of test object m toward the source object M in Figure 1b. However, gravity in Newtonian perspective is an interaction force between massive objects, therefore it can start the motion from rest. From the point of view of Figure 1b, even if we accept a massive source curves the space time, there is not any reason to attract a test object toward left side. In another word, there is no reason for the curvature to start the motion of a test particle. We may accept there is a curvature, but this does not require a motion to start. When we assume a curvature leads to an attraction and motion in fact we still become under the effect of traditional gravity concept because we perceive the below side like the direction of gravity as on earth for Figure 1a, but in Figure 1b terminates this perceive.

How can the curvature be responsible from the gravity and motion due to gravity inferred from the curvature? When we assume there places a test object on the horizon in Figure 1b and it is at rest, how can curvature lead to a motion for this test particle toward the left side? Is there a force toward the curvature? We traditionally expect a force toward the curvature although curvature theory rejects the force concept. So, how can a curvature behave as a force. Assume a parabola on the sheet in Figure 1b, how can we say if we place a point object m at the starting point of the parabola, it should move on the parabola? Can geometry solely force the particle to move? We intuitively invoke an interaction force directed toward the curvature although we terminate the force concept by introducing the curvature concept.

Another important question is that what will happen if we take another test object from the left side of the source in Figure 1b? Will not the source attract the test object because the curvature is toward left side in figure? Or, is there another curvature toward right side with respect to the new test object? If so, what kind of curvature is that? Or, what kind of space is that? How can a space be curved through in every direction when we place a mass M on itself?

To understand the last question better let us imagine a 1-Dimensional system as in Figure 1c. A 1-Dimensional source mass M and a 1-Dimensional test mass m with a rod shape are placed to a 1-Dimensional space. Then, curvature corresponds to a downward strain on the 1-Dimensional space. If we accept the curvature in the strain form can attract the test object toward down side, what happens if we place another 1-Dimensional test particle below the source particle? Because we accept the strain curvature is directed toward down side, what will the below test particle happen? Shall we say it will not be attracted, or it will be attracted upward direction? If we say the latter, this means there is another curvature toward above side but this is impossible because the strain curvature is toward downward. On the other hand, how can a strain curvature start the motion as considered in Figure 1d. It makes no sense that a horizontal strain curvature leads to a motion on a test object as considered a gravitational attraction. Obviously, we need a force interpretation for gravity other than a curvature interpretation.

We should stop being single minded, and thinking the traditionally accepted interpretations are concrete truths about the nature. We should re-think the curvature interpretation of general relativity. Is it just a mathematical tool and model to predict the phenomena in nature, or is it the concrete phenomenon in nature? According to the above criticisms it seems that the curvature interpretation is just a useful mathematical tool to predict the phenomena in nature, rather than the phenomenon itself.

Received April 28, 2020; Accepted May 10, 2020

References

Dil, E. 2019, Prespacetime Journal, 10, 1119
Friedman, Y. 2016, Europhysics Letters, 116
Friedman, Y., Steiner, J. 2016, Europhysics Letters, 113
Friedman, Y., Steiner, J. 2017, Europhysics Letters, 117

ISSN: 2153-8301   Prespacetime Journal   www.prespacetime.com
Published by QuantumDream, Inc.

Exploration

Derivation of the θ-parameter in Quantum Chromodynamics

Ervin Goldfain[*]

Advanced Technology and Sensor Group, Welch Allyn Inc., Skaneateles Falls, NY 13153

Abstract

An ongoing challenge of the Standard Model is to explain the nearly vanishing magnitude of CP symmetry breaking in Quantum Chromodynamics (QCD). This challenge goes by the name of the *strong CP problem* and is quantified by the θ-parameter of the QCD Lagrangian. It is known that *axions* are hypothetical particles conjectured to offset the contribution of the θ-parameter and restore the CP symmetry of QCD. Starting from the near equilibrium interpretation of gluon dynamics, here we bypass the axion conjecture and derive the numerical value of the θ-parameter in close agreement with experimental bounds. A surprising finding of this brief report is that the strong CP and the baryon asymmetry problems appear to be related to each other.

Keywords: Quantum Chromodynamics, strong CP problem, axions, fractional dynamics, non-equilibrium field theory.

The violation of parity (P) and time-reversal (T) symmetries in QCD follows from the nontrivial structure of the quantum vacuum and is embodied in the so-called θ-term entering the Lagrangian [1-3, 10]

$$\Delta L_{QCD} = L_{QCD} - L_{QCD}^{(\theta=0)} = \theta \frac{g_s^2}{64\pi^2} G_{\mu\nu}^a \overline{G}^{a\mu\nu} \tag{1}$$

in which g_s is the strong coupling charge, $G_{\mu\nu}^a$ the gluon field strength and $\overline{G}^{a\mu\nu}$ its dual. The θ-term can be written as

$$G_{\mu\nu}^a \overline{G}^{a\mu\nu} = \partial_\mu K^\mu \tag{2}$$

where K_μ represents a current built from the components of the gluon field,

---

[*]Correspondence: Ervin Goldfain, Ph.D., Photonics CoE, Welch Allyn Inc., Skaneateles Falls, NY 13153, USA
E-mail: ervingoldfain@gmail.com

$$\mathbf{A}_{\mu} = A_{\mu}^{a}\frac{\lambda^{a}}{2} \tag{3}$$

Consider a gluon field configuration that starts off at $t = -\infty$ and ends up at $t = +\infty$. It can be shown that the following integral is non-vanishing

$$\frac{g_{s}^{2}}{64\pi^{2}}\int d^{4}x\, G_{\mu\nu}^{a}\overline{G}^{a\mu\nu} \neq 0 \tag{4}$$

In particular [3]

$$\frac{g_{s}^{2}}{64\pi^{2}}\int d^{4}x\, G_{\mu\nu}^{a}\overline{G}^{a\mu\nu} = \frac{g_{s}^{2}}{64\pi^{2}}\int d^{3}x\, K_{0}\bigg|_{t=-\infty}^{t=+\infty} = 1 \tag{5}$$

We proceed below with the assumption that the dynamics of the gluon field is in *near equilibrium* conditions and that the integral (5) can be well approximated by

$$\frac{g_{s}^{2}}{64\pi^{2}}\left\langle G_{\mu\nu}^{a}\overline{G}^{a\mu\nu}\right\rangle \cdot V_{4} = 1 \Rightarrow \left\langle G_{\mu\nu}^{a}\overline{G}^{a\mu\nu}\right\rangle = O(V_{4}^{-1}) \tag{6}$$

where V_{4} is the finite four-dimensional integration volume, commensurate with the QCD scale via

$$V_{4} = \int d^{4}x = O(\Lambda_{QCD}^{-4}) \tag{7}$$

Because the product of a field Lagrangian with the four-dimensional volume defining its range yields the field action, by (1), (6) and (7) we obtain

$$\theta = \left\langle \Delta L_{QCD}\right\rangle \cdot V_{4} = \left\langle S_{QCD} - S_{QCD}^{(\theta=0)}\right\rangle \tag{8}$$

It is known that the field action can be specified up to an arbitrary constant [7]. Since gluon dynamics is considered in near equilibrium conditions, we choose an infinitesimal departure of the action from its equilibrium value S_{ech} and rewrite (8) as

$$\theta = \left\langle (S_{QCD} - S_{ech}) - (S_{QCD}^{(\theta=0)} - S_{ech})\right\rangle = \left\langle \delta S_{QCD} - \delta S_{QCD}^{(\theta=0)}\right\rangle \tag{9}$$

Appealing to [4-6] motivates the plausible assumption that both δS_{QCD} and $\delta S_{QCD}^{(\theta=0)}$ scale linearly with the dimensional deviation from four spacetime dimensions described by $\varepsilon = 4 - D << 1$. As a result, (9) leads to

$$\theta = O(\varepsilon) - O(\varepsilon) = O(\varepsilon^2) \tag{10}$$

Furthermore, since ε flows with the energy scale, it likely reaches its uppermost observable value close to the formation of the cosmic microwave background (CMB) [8]. It is therefore reasonable to conjecture that the maximal dimensional deviation is given by

$$\varepsilon_{max} = O(10^{-5}) \tag{11}$$

We recall that this scenario falls in line with the requirements set by the Sakharov conditions for baryogenesis. Given the long-range effects carried by dimensional deviation ε throughout all energy scales, it makes sense to assume that (11) persists as a relic effect at the Standard Model scale. With these considerations in mind and replacing (11) in (10) yields

$$\boxed{\theta \leq O(\varepsilon_{max}^2) = O(10^{-10})} \tag{12}$$

in close agreement with current experimental data on the neutron dipole moment [9-10].

Received February 22, 2020; Accepted March 30, 2020

References

1. Cheng, T. P., Li, L.F., "Gauge theory of elementary particle physics", Oxford Univ. Press, 1984.

2. Kaku, M., "Quantum Field Theory, A Modern Introduction", Oxford Univ. Press, 1993.

3. Donoghue, J. F. *et al.,* "Dynamics of the Standard Model", Cambridge Univ. Press, 1992.

4. Available at the following sites:
http://www.aracneeditrice.it/aracneweb/index.php/pubblicazione.html?item=9788854889972
https://www.researchgate.net/publication/278849474_Introduction_to_Fractional_Field_Theory_consolidated_version

5. http://www.ejtp.info/articles/ejtpv7i24p219.pdf

6. Available at the following site:
https://www.researchgate.net/publication/266578650_REFLECTIONS_ON_THE_FUTURE_OF_PARTICLE_THEORY

7. see, e.g. A. O. Barut, "Electrodynamics and Classical Theory of Fields and Particles", Dover Publications, 1980.

8. Available at the following site:
https://www.researchgate.net/publication/338698237_Baryon_Asymmetry_from_the_Minimal_Fractal_Manifold

9) Available at the following site:
https://journals.aps.org/prl/accepted/b607fY80Z3a12a6ab8689246ed949444cd5500f42

10. http://cds.cern.ch/record/560630/files/0206123.pdf

Perspective

Towards Gross-Pitaevskiian Description of Solar System & Galaxies

Victor Christianto[1*], Florentin Smarandache[2] & Yunita Umniyati[3]

[1]Malang Institute of Agriculture (IPM), Malang, Indonesia
[2]Dept. of Math. Sci., Univ. of New Mexico, Gallup, USA
[3]Dept. Mechatronics, Swiss-German Univ., Tangerang, Indonesia

Abstract

In this paper, we argue that Gross-Pitaevskii model can be a more complete description of both solar system and spiral galaxies, especially taking into account the nature of chirality and vortices in galaxies. We also hope to bring out some correspondence among existing models, e.g., the topological vortex approach, Burgers equation in the light of KAM theory, and the Cantorian Navier-Stokes approach. We hope further investigation can be done around this line of approach.

Keywords: Solar system, galaxy, Gross-Pitaevskii, Burgers equation, Navier-Stokes equation.

1. Introduction

From time to time, astronomy and astrophysics discoveries have opened our eyes that the Universe is much more complicated than what it seemed in 100-200 years ago. And despite all pervading popularity of General Relativistic extension to Cosmology, it seems still worthy to remind us to old concepts of Cosmos, for instance the *Hydor theory* of Thales ("that water is the essential element in the Cosmos")[1], and also Heracleitus ("ta *panta rhei kai ouden menei*").[2] Therefore, we can ask: does it mean that the Ultimate theory that we try to find should correspond to hydrodynamics or some kind of turbulence theory?

An indicator of complex turbulence phenomena in Our Universe is the Web like structure. The *Cosmic Web* is the fundamental spatial organization of matter on scales of a few up to a hundred Megaparsec. Galaxies and intergalactic gas matter exist in a wispy web-like arrangement of dense compact clusters, elongated filaments, and sheet-like walls, amidst large near-empty void regions. The filaments are the transport channels along which matter and galaxies flow into massive high-density cluster located at the nodes of the web. The web-like network is shaped by the tidal force field accompanying the inhomogeneous matter distribution.

---

*Correspondence: Victor Christianto, Malang Institute of Agriculture (IPM), Malang, Indonesia.
Email: victorchristianto@gmail.com

[1] https://www.philosophy.gr/presocratics/thales.htm
[2] https://carinawestling.wordpress.com/2010/12/03/τὰ-πάντα-ῥεῖ-καὶ-οὐδὲν-μένει-ta-panta-rhei-kai-ouden-menei/

May be part of that reason that in recent years, there is growing interest to describe the Universe we live in from the perspective of scale-invariant turbulence approach. Such an approach is not limited to hydrodynamics Universe model a la Gibson & Schild, but also from Kolmogorov turbulence approach as well as from String theory approach (some researchers began to explore String-Turbulence).

Recently, Pitkanen describes a solar system model inspired by spiral galaxies [1-2]. While we appreciate his new approach, we find it lacks discussion on the nature of vortices and chirality in galaxy.

In this article, we show some correspondences among existing models, so we discuss shortly, the topological vortice approach, Burgers equation in the light of KAM theory and Golden Mean, and the Cantorian Navier-Stokes approach. We will point out how vortices, turbulence and chirality nature of galaxies seem to suggest a quantised vortex approach, which in turn it corresponds to Gross-Pitaevskiian description.

2. Quantised vortices approach (see also ref. [42-43])

Here we present Bohr-Sommerfeld quantization rules for planetary orbit distances, which results in a good quantitative description of planetary orbit distance in the solar system [6][6b][7]. Then we find an expression which relates the torsion vector and quantized vortices from the viewpoint of Bohr-Sommerfeld quantization rules [3]. Further observation of the proposed quantized vortices of superfluid helium in astro-physical objects is recommended.

<u>Bohr-Sommerfeld quantization rules and quantized vortices</u>

Sonin's book [42] can be paraphrased as follows:

> The movement of vortices has been a region of study for over a century. During the old style time of vortex elements, from the late 1800s, many fascinating properties of vortices were found, starting with the outstanding Kelvin waves engendering along a disconnected vortex line (Thompson, 1880). The primary object of hypothetical investigations around then was a dissipationless immaculate fluid (Lamb, 1997). It was difficult for the hypothesis to find a shared opinion with try since any old style fluid shows gooey impacts. The circumstance changed after crafted by Onsager (1949) and Feynman (1955) who uncovered that turning superfluids are strung by a variety of vortex lines with quantised dissemination. With this revelation, the quantum time of vortex elements started.

The quantization of circulation for nonrelativistic superfluid is given by [3]:

$$\oint v\,dr = N\frac{\hbar}{m_s} \tag{1}$$

where $N, \hbar, m_s$ represents winding number, reduced Planck constant, and superfluid particle's mass, respectively [3]. And the total number of vortices is given by [44]:

$$N = \frac{\omega . 2\pi r^2 m}{\hbar} \qquad (2)$$

And based on the above equation (2), Sivaram & Arun [44] are able to give an estimate of the number of galaxies in the universe, along with an estimate of the number stars in a galaxy.

However, they do not give explanation between the quantization of circulation (3) and the quantization of angular momentum. According to Fischer [3], the quantization of angular momentum is a relativistic extension of quantization of circulation, and therefore it yields Bohr-Sommerfeld quantization rules.

Furthermore, it was suggested in [6] and [7] that Bohr-Sommerfeld quantization rules can yield an explanation of planetary orbit distances of the solar system and exoplanets. Here, we begin with Bohr-Sommerfeld's conjecture of quantization of angular momentum. As we know, for the wavefunction to be well defined and unique, the momenta must satisfy Bohr-Sommerfeld's quantization condition:

$$\oint_{\Gamma} p.dx = 2\pi.n\hbar, \qquad (3)$$

for any closed classical orbit Γ. For the free particle of unit mass on the unit sphere the left-hand side is:

$$\int_0^T v^2 .d\tau = \omega^2 .T = 2\pi.\omega, \qquad (4)$$

where $T = \dfrac{2\pi}{\omega}$ is the period of the orbit. Hence the quantization rule amounts to quantization of the rotation frequency (the angular momentum): $\omega = n\hbar$. Then we can write the force balance relation of Newton's equation of motion:

$$\frac{GMm}{r^2} = \frac{mv^2}{r}. \qquad (5)$$

Using Bohr-Sommerfeld's hypothesis of quantization of angular momentum (4), a new constant g was introduced:

$$mvr = \frac{ng}{2\pi}. \qquad (6)$$

Just like in the elementary Bohr theory (just before Schrodinger), this pair of equations yields a known simple solution for the orbit radius for any quantum number of the form:

$$r = \frac{n^2 . g^2}{4\pi^2 . GMm^2},$$
(7)

or

$$r = \frac{n^2 . GM}{v_o^2},$$
(8)

where r, n, G, M, v_o represents orbit radii (semimajor axes), quantum number (n=1,2,3,…), Newton gravitation constant, and mass of the nucleus of orbit, and specific velocity, respectively. In equation (10), we denote:

$$v_0 = \frac{2\pi}{g} GMm.$$
(9)

The value of m and g in equation (9) are adjustable parameters.

Interestingly, we can remark here that equation (8) is exactly the same with what is obtained by Nottale using his Schrödinger-Newton formula [8]. Therefore here we can verify that the result is the same, either one uses Bohr-Sommerfeld quantization rules or Schrödinger-Newton equation. The applicability of equation (8) includes that one can predict new *exoplanets* (i.e., extrasolar planets) with remarkable result.

Therefore, one can find a neat correspondence between Bohr-Sommerfeld quantization rules and motion of quantized vortices in condensed-matter systems, especially in superfluid helium [3]. Here we propose a conjecture that superfluid vortices quantization rules also provide a good description for the motion of galaxies, especially with respect to their chirality nature, as will be discussed later.

3. Golden ratio is directly related to KAM turbulence via Burgers equation

The Cosmic Web is the fundamental spatial organization of matter on scales of a few up to a hundred Megaparsec. Galaxies and intergalactic gas matter exist in a wispy weblike arrangement of dense compact clusters, elongated filaments, and sheetlike walls, amidst large near-empty void regions. The filaments are the transport channels along which matter and galaxies flow into massive high-density cluster located at the nodes of the web. The weblike network is shaped by the tidal force field accompanying the inhomogeneous matter distribution.[15]

Structure in the Universe has risen out of tiny primordial (Gaussian) density and velocity perturbations by means of gravitational instability. The large-scale anisotropic force field induces anisotropic gravitational collapse, resulting in the emergence of elongated or flattened matter configurations. The simplest model that describes the emergence of structure and complex patterns in the Universe is the Zeldovich Approximation (ZA).[15]

It is our hope that the new approach of CA Adhesion model of the Universe can be verified either with lab experiments, computer simulation, or by large-scale astronomy observation data.

From Zeldovich Approximation to Burgers' equation to Cellular Automaton model

In this section, we will outline a route from ZA to Burgers' equation and then to CA model. The simplest model that describes the emergence of structure and complex patterns in the Universe is the Zeldovich Approximation (ZA). In essence, it describes a ballistic flow, driven by a constant (gravitational) potential. The resulting Eulerian position x(t) at some cosmic epoch t is specified by the expression[15]:

$$x(t) = q + D(t)u_o(q), \tag{10}$$

where q is the initial "Lagrangian" position of a particle, D(t) the time-dependent structure growth factor and

$$u_0 = -\nabla_q \Phi_0 \tag{11}$$

its velocity. The nature of this approximation may be appreciated by the corresponding source-free equation of motion,

$$\frac{\partial u}{\partial D} + (u \cdot \nabla_x)u = 0. \tag{12}$$

The use of ZA is ubiquitous in cosmology. One major application is its key role in setting up initial conditions in cosmological N-body simulations. Of importance here is its nonlinear extension in terms of *Adhesion Model* [15]:

The ZA breaks down as soon as self-gravity of the forming structures becomes important. To 'simulate' the effects of self-gravity, Gurbatov *et al.* included an artificial viscosity. This results in the Burgers' equation as follows [15]:

$$\frac{\partial u}{\partial D} + (u \cdot \nabla_x)u = v.\nabla_x^2 u, \tag{13}$$

a well known PDE from fluid mechanics. This equation has an exact analytical solution, which in the limit of $v \to 0$, the solution is [15]:

$$\phi(x, D) = \max_q \left[\Phi_0(q) - \frac{(x-q)^2}{2D} \right]. \tag{14}$$

This leads to a geometric interpretation of the Adhesion Model. The solution follows from the evaluation of the convex hull of the velocity potential modified by a quadratic term. We found that the solution can also be found by computing the weighted Voronoi diagram of a mesh weighted with the velocity potential. For more detailed discussion on Adhesion Model of the Universe, see for example [18].

Now, let us consider another routes to solve Burgers equation: (a) by numerical computation with *Mathematica*, see [17]; and (b) by virtue of CA approach. Let us skip route (a), and discuss less known approach of cellular automata.

We start with the Burgers' equation with Gaussian white noise which can be rewritten as follows [16]:

$$\frac{\partial u}{\partial t} + \xi = 2u\frac{\partial u}{\partial x} + \frac{\partial^2 u}{\partial x^2} + \eta. \tag{15}$$

By introducing new variables and after straightforward calculations, we have the automata rule [16]:

$$\phi_i^{t+1} = \phi_{i-1}^t + \max[0, \phi_i^t - A, \phi_i^t + \phi_{i+1}^t - B, \Psi_i^t - \phi_{i-1}^t]$$
$$- \max[0, \phi_{i-1}^t - A, \phi_{i-1}^t + \phi_i^t - B, \Phi_i^t + \phi_{i-1}^t] \tag{16}$$

In other words, in this section we give an outline of a plausible route from ZA to Burgers' equation then to CA model, which suggests that it appears possible –at least in theory- to consider a nonlinear cosmology based on CA Adhesion model.

From KAM theory to Golden section

Another possible way to describe the complex structure of Universe, is the Kolmogorov-Arnold-Moser (KAM) theorem, which states that if the system is subjected to a weak nonlinear perturbation, some of the invariant tori are deformed and survive, while others are destroyed. The ones that survive are those that have "sufficiently irrational frequencies" (the non-resonance condition, so they do not interfere with one another). The golden ratio being the most irrational number is often evident in such systems of oscillators. It is also physically significant in that circles with golden mean frequencies are the last to break up in a perturbed dynamical system, so the motion continues to be quasi-periodic, i.e., recurrent but not strictly periodic or predictable.

An important consequence of the KAM theorem is that for a large set of initial conditions, the motion remains perpetually quasi-periodic, and hence stable. KAM theory has been extended to non-Hamiltonian systems and to systems with fast and slow frequencies.

Those KAM tori that are not destroyed by perturbation become invariant Cantor sets, or "Cantori". The frequencies of the invariant Cantori approximate the golden ratio. The golden ratio effectively enables multiple oscillators within a complex system to co-exist without blowing up the system. But it also leaves the oscillators within the system free to interact globally (by resonance), as observed in the coherence potentials that turn up frequently when the brain is processing information.

Obviously, this can be tied in to the creation of subatomic particles such as electrons and positrons. At a certain scale of smallness, the media in the local volume becomes isotropic, while larger volumes exhibit occupation by ever-larger turbulence formations and exhibit extremes of **an**isotropy in the media.

The Kolmogorov Limit is $10e^{-58}$ m, which is the smallest vortex that can exist in the aether media. Entities smaller than this, down to the SubQuantum infinitesimals (Bhutatmas) (vortex lines) are the primary cause of gravitation (a "sink" model of gravitation caused by superluminal infinitesimals).[3]

Figure 1.   Turbulent flow generated by the tip vortex of the aeroplane wing shown up by red agricultural dye. (after Mae Wan-Ho [38]).

Shadow gravity is valid in the situation of gravitational interaction between two discrete masses that divert the ambient gravitational flux-density away from each other. This happens due to absorption (rare), scattering (more common), and refraction (most of the time) of gravitational infinitesimals.

Gravitational flux density is a variable depending on stellar, interstellar, and intergalactic events. A simplified model of vorticity fields in large scale structures of the Universe is depicted below:

What is more interesting here, is that it can be shown that there is correspondence between Golden section and in coupled oscillators and KAM Theorem, but also between Golden section and Burgers equation. [35]. For more discussion, on Golden Mean and its ramifications, see for instance [39][40][41].

---

[3] Thanks to discussions with Robert Neil Boyd, PhD.

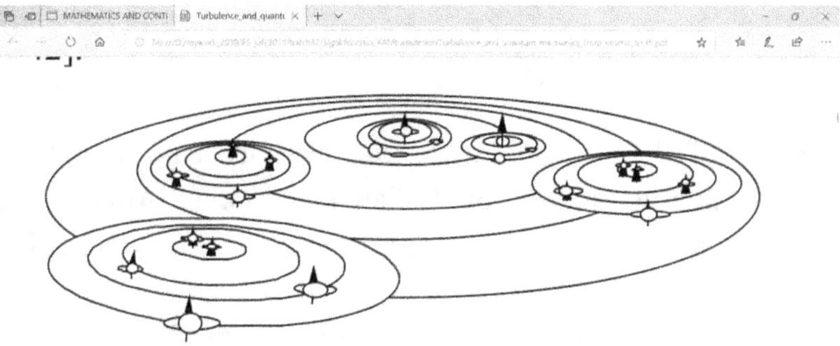

Fig.2 Description of internal (iso-spin) versus external vorticity fields in cosmology [41].

Figure 2. Vorticity fields in cosmology (after Siavash Sohrab [34]).

4. Cantorian Navier-Stokes approach

Vorticity as the driver of Accelerated Expansion

According to Ildus Nurgaliev [26], velocity vector V_α of the material point is projected onto coordinate space by the tensor of the second rank $H_{\alpha\beta}$:

$$V_\alpha = H_{\alpha\beta} R^\beta \tag{17}$$

where the Hubble matrix can be defined as follows for a homogeneous and isotropic universe:

$$H_{\alpha\beta} = \begin{pmatrix} H & \pm\omega & \pm\omega \\ \mp\omega & H & \pm\omega \\ \mp\omega & \mp\omega & H \end{pmatrix} \tag{18}$$

where the global average vorticity may be zero, though not necessarily [7]. Here the Hubble law is extended to 3x3 matrix.

Now we will use Newtonian equations to emphasize that cosmological singularity is consequence of the too simple model of the flow, and has nothing to do with special or general relativity as a cause [26]. Standard equations of Newtonian hydrodynamics in standard notations read:

$$\frac{d\vec{\upsilon}}{dt} = \frac{\partial\vec{\upsilon}}{\partial t} + \vec{\upsilon}\nabla\vec{\upsilon} = -\nabla\varphi + \frac{1}{\rho}\nabla\rho + \frac{\mu}{\rho}\Delta\vec{\upsilon} + ..., \tag{19}$$

$$\frac{\partial \rho}{\partial t} + \nabla \rho \vec{\upsilon} = 0, \tag{20}$$

$$\Delta \varphi = 4\pi G \rho \tag{21}$$

Procedure of separating of diagonal H, trace-free symmetrical σ, and anti-symmetrical ω elements of velocity gradient was used by Indian theoretician Amal Kumar Raychaudhury (1923-2005). The equation for expansion θ, sum of the diagonal elements of [7]:

$$\dot{\theta} + \frac{1}{3}\theta^2 + \sigma^2 - \omega^2 = -4\pi G\rho + div(\frac{1}{\rho}\sum f) \tag{22}$$

is most instrumental in the analysis of singularity and bears the name of its author. [26]

System of (25)-(27) gets simplified up to two equations [26]:

$$\dot{\theta} + \frac{1}{3}\theta^2 - \omega^2 = 0, \tag{23}$$

$$\dot{\omega} + \frac{2}{3}\theta\omega = 0. \tag{24}$$

Recalling $\theta = 3H$, the integral of (30) takes the form [26]:

$$H^2 = H_\infty^2 - \frac{3\omega_0^2 R_0^4}{R^4}. \tag{25}$$

<u>How to write down Navier-Stokes equations on Cantor Sets</u>

Now we can extend further the Navier-Stokes equations to Cantor Sets, by keeping in mind their possible applications in cosmology.

By defining some operators as follows:

1. In Cantor coordinates [28]:

$$\nabla^\alpha \cdot u = div^\alpha u = \frac{\partial^\alpha u_1}{\partial x_1^\alpha} + \frac{\partial^\alpha u_2}{\partial x_2^\alpha} + \frac{\partial^\alpha u_3}{\partial x_3^\alpha}, \tag{26}$$

$$\nabla^\alpha \times u = curl^\alpha u = \left(\frac{\partial^\alpha u_3}{\partial x_2^\alpha} - \frac{\partial^\alpha u_2}{\partial x_3^\alpha}\right)e_1^\alpha + \left(\frac{\partial^\alpha u_1}{\partial x_3^\alpha} - \frac{\partial^\alpha u_3}{\partial x_1^\alpha}\right)e_2^\alpha + \left(\frac{\partial^\alpha u_2}{\partial x_1^\alpha} - \frac{\partial^\alpha u_1}{\partial x_2^\alpha}\right)e_3^\alpha. \tag{27}$$

2. In Cantor-type cylindrical coordinates [29, p.4]:

$$\nabla^\alpha \cdot r = \frac{\partial^\alpha r_R}{\partial R^\alpha} + \frac{1}{R^\alpha}\frac{\partial^\alpha r_\theta}{\partial \theta^\alpha} + \frac{r_R}{R^\alpha} + \frac{\partial^\alpha r_z}{\partial z^\alpha}, \tag{28}$$

$$\nabla^{\alpha} \times r = \left(\frac{1}{R^{\alpha}} \frac{\partial^{\alpha} r_{\theta}}{\partial \theta^{\alpha}} - \frac{\partial^{\alpha} r_{\theta}}{\partial z^{\alpha}} \right) e_{R}^{\alpha} + \left(\frac{\partial^{\alpha} r_{R}}{\partial z^{\alpha}} - \frac{\partial^{\alpha} r_{z}}{\partial R^{\alpha}} \right) e_{\theta}^{\alpha} + \left(\frac{\partial^{\alpha} r_{\theta}}{\partial R^{\alpha}} + \frac{r_{R}}{R^{\alpha}} - \frac{1}{R^{\alpha}} \frac{\partial^{\alpha} r_{R}}{\partial \theta^{\alpha}} \right) e_{z}^{\alpha} . \tag{29}$$

Then Yang, Baleanu and Machado are able to obtain a general form of the Navier-Stokes equations on Cantor Sets as follows [28, p.6]:

$$\rho \frac{D^{\alpha} \upsilon}{Dt^{\alpha}} = -\nabla^{\alpha} \cdot (pI) + \nabla^{\alpha} \left[2\mu \left(\nabla^{\alpha} \cdot \upsilon + \upsilon \cdot \nabla^{\alpha} \right) - \frac{2}{3} \mu \left(\nabla^{\alpha} \cdot \upsilon \right) I \right] + \rho b \tag{30}$$

The next task is how to find observational cosmology and astrophysical implications. This will be the subject of future research.

5. Correspondence with Gross-Pitaevskiian description and a description of chirality nature of galaxies

In this section we will point out how vortices and chirality nature of galaxies seem to suggest a superfluid vortex approach, which in turn it corresponds to Gross-Pitaevskiian description. The nature and origin of chirality in galaxies remain an elusive topic to explain. However we can recall some recent works to suggest an explanation.

First of all, let us quote from abstract of a recent paper [9], where Tapio Simula wrote, which can be rephrased as follows:

> Right now, and electromagnetism have a similar starting point and are new properties of the superfluid universe, which itself rises up out of the hidden aggregate structure of progressively basic particles, for example, atoms. The Bose–Einstein condensate is identified as the tricky dull matter of the superfluid universe with vortices and phonons, separately, comparing to huge charged particles and massless photons.

In lieu of his model of electromagnetic and gravitation fields in terms of superfluid vortices, we can also come up with a model of chirality in cosmology from Proca equations. As Proca equations can be used to describe electromagnetic field of superconductor, we find it as a possible approach too [45].

Now we are going to discuss how it can be used as a model of chirality nature of galaxies. Cappoziello and Lattanzi argue that spiral galaxies are axi-symmetric objects showing 2D-chirality when projected onto a plane [36]. In their enantiomers model, chirality in spiral galaxies and chirality Spiral galaxies are axi-symmetric objects showing 2D-chirality when projected onto a plane, and their progressive loss of chirality surrounding its galaxy center, can point out of vorticity in superfluidity.

See the following figure, which is quite in agreement with Figure 2 by Sohrab:

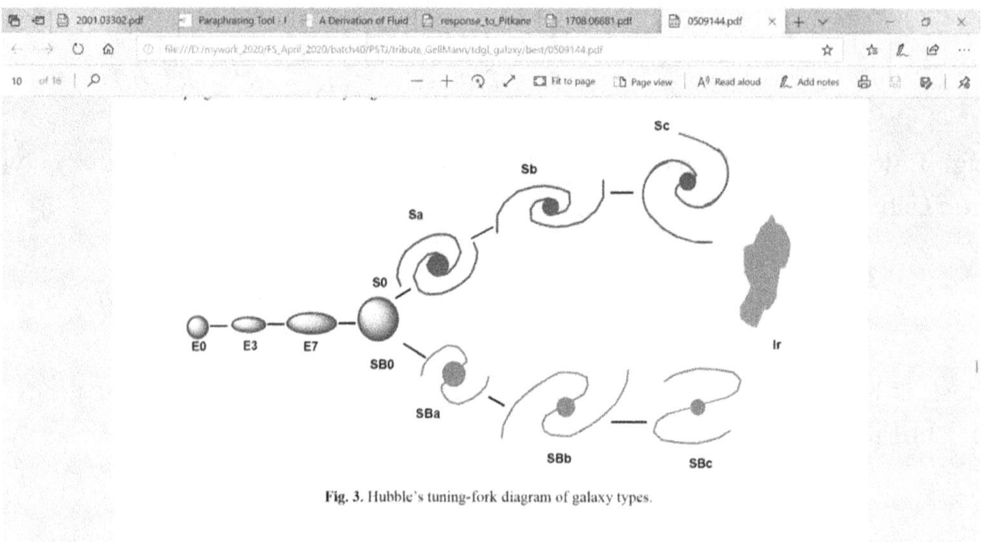

Figure 3 . After Cappoziello & Lattanzi [36].

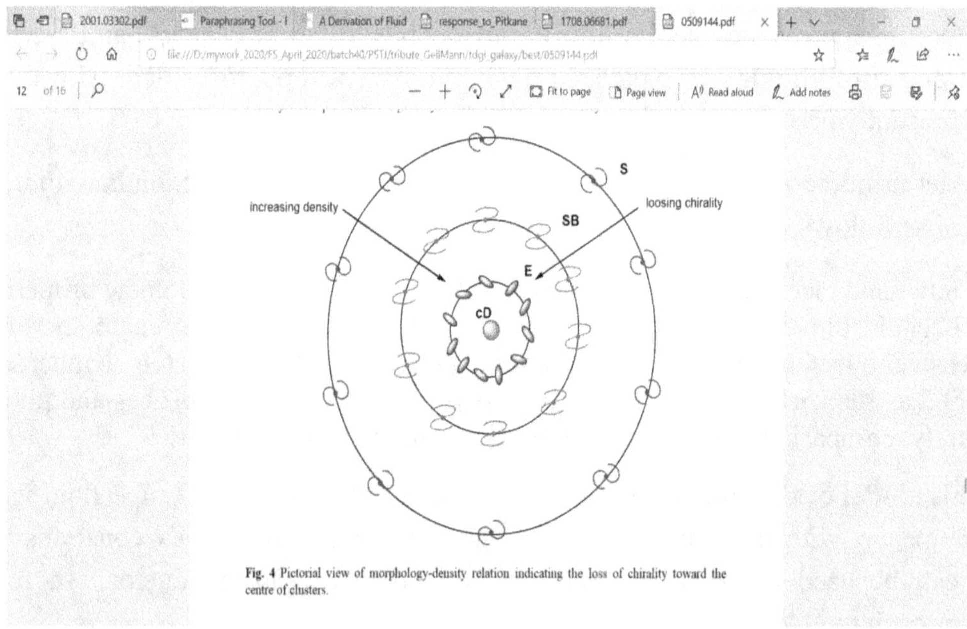

Figure 4. Progressive loss of chirality. After Cappoziello & Lattanzi [36].

With regards to question posed above: what kind of medium of interaction capable of doing such a quantal action? Allow us to quote from Fernandez-Hernandez *et al's* abstract [37], which can be paraphrased as follows:

> The ultra-light scalar fields are likewise called scalar field dull issue model. Right now study turn bends for low surface brilliance winding cosmic systems utilizing two scalar

field models: the Gross-Pitaevskii *Bose-Einstein condensate in the Thomas-Fermi estimation* and a scalar field arrangement of the Klein-Gordon condition. We additionally utilized the zero circle guess universe model where photometric information isn't thought of.

Therefore, we come up with a conclusion that it seems *Gross-Pitaevskiian description of Bose-Einstein condensate is necessary for correct modelling of spiral and non-spiral galaxies.* Interestingly, in two rather old papers both of us (VC & FS) have argued for derivation of Schrödinger equation model of planetary orbits of solar system from TDGL (Gross-Pitaevskii) description [46-47]. Therefore, it seems we can arrive at this conclusion: it is possible to come up with consistent description of both solar system and galaxy dynamics, including its chirality and rotation curves, by virtue of Gross-Pitaevkiian description of BEC/superfluidity.

6. Conclusion

In this paper, we have discussed several approaches in description of planetary systems as well as galaxies. Sections 2-4 have been presented in earlier papers.

Therefore, we come up with a conclusion that there are sufficient grounds to argue in favour of Gross-Pitaevskiian description of Bose-Einstein condensate; i.e. it is necessary for correct modelling of spiral and non-spiral galaxies. Interestingly, in a rather old paper both of us (VC & FS) have argued for derivation of Schrodinger equation model of planetary orbits of solar system from TDGL (Gross-Pitaevskii) description.

Summarizing, it seems we can arrive at this conclusion: it is possible to come up with consistent description of both solar system and galaxy dynamics, *including its chirality and rotation curves, by virtue of Gross-Pitaevkiian description of BEC.*

Of course, this short article is far from being complete. We hope further investigation can be done around this line of approach. The remaining questions include how to find observational cosmology and astrophysical implications. Future research is recommended.

Received April 17, 2020; Accepted May 17, 2020

References

[1] M. Pitkanen. Modeling of Solar System as a Miniature Version of Spiral Galaxy. *Prespacetime J.,* April 2020, Vol. 11, Issue 2, pp. 109-121

[2] M. Pitkanen. A Model for the Formation of Galaxies. *Prespacetime J.,* April 2020, Vol. 11, Issue 2, pp. 90-99.

[3] Fischer, U., (1999) Motion of quantized vortices as elementary objects, *Ann. Phys.* (N.Y.) 278, 62-85, and also in arXiv:cond-mat/9907457

[4] Bamba, K., Capozziello, C., Nojiri, S. & S.D. Odintsov (2012) Dark energy cosmology: the equivalent description via different theoretical models and cosmography tests, arXiv:1205.3421 [gr-qc] see p. 94.

[5] Starkman, G.D. (2012) Modifying gravity: You can't always get what you want, arXiv: 1201.1697 [gr-qc].

[6] Christianto, V. (2006) On the origin of macroquantization in astrophysics and celestial motion, *Annales de la Fondation Louis de Broglie,* Volume 31 no 1; [6b] F. Smarandache & Christianto, V. (2006) Schrodinger equation and the quantization of celestial systems. *Progress in Physics*, Vol. 2, April 2006.

[7] Christianto, V., (2004) A Cantorian superfluid vortex and the quantization of planetary motion, *Apeiron*, Vol. 11, No. 1, January 2004, http://redshift.vif.com

[8] Nottale, L., *Astron. Astrophys.* **327,** 867-889 (1997).

[9] T. Simula. Gravitational Vortex Mass in a Superfluid. arXiv: 2001.03302 (2020).

[10] Wang, X-S. (2005) Derivation of Newton's Law of Gravitation Based on a Fluid Mechanical Singularity Model of Particles, arXiv:physics/0506062.

[11] Wang, X-S. (2006) Derivation of the Schrodinger equation from Newton's Second Law Based on a Fluidic Continuum Model of Vacuum and a Sink Model of Particles, arXiv:physics/0610224.

[12] Silk, J. (2001) The formation of galaxy disks, arXiv:astro-ph/0010624

[13] Kain, B., & H.Y. Ling (2010) Vortices in Bose-Einstein condensate dark matter, arXiv:1004.4692 [hep-ph]

[14] Brook, M.N. (2010) *Cosmology meets condensed matter.* PhD Dissertation,The University of Nottingham. 171 p.

[15] J. Hidding, R. van de Weygaert, G. Vegter, B.J.T. Jones, M. Teillaud. The Sticky Geometry of the Cosmic Web. SCG'12, June 17–20, 2012, Chapel Hill, North Carolina, USA. ACM 978-1-4503-1299-8/12/06. arXiv: 1205.1669 [astro-ph.CO] (2012); [15a] J. Hidding, S.Shandarin, R. van de Weygaert. The Zeldovich Approximation: key to understanding Cosmic Web complexity. *Mon. Not. Royal Astron.* Soc. 1-37 (2013)

[16] Xin-She Yang & Y. Young. Cellular Automata, PDEs, and Pattern Formation. arXiv: 1003.1983 (2010)

[17] Richard H. Enns & George C. McGuire. *Nonlinear Physics with Mathematica for Scientists and Engineers.* Boston: Birkhäuser, 2001. See pp. 314-316.

[18] J. Hidding. *Adhesion: a sticky way of understanding Large Scale Structure.* 2010. 180 p.

[19] O. Hahn. Collisionless Dynamics and the Cosmic Web, a chapter in R. van de Weygaert, S. Shandarin, E. Saar & J. Einasto, eds. *The Zeldovich Universe, Proceedings IAU Symposium No. 308,* 2014. Also in arXiv: 1412.5197 [astro-ph.CO]

[20] G. Argentini, Exact solution of a differential problem in analytical fluid dynamics, arXiv:math.CA/0606723 (2006).

[21] Victor Christianto & Florentin Smarandache. "An Exact Mapping from Navier Stokes equation to Schrödinger equation via Riccati equation." *Progress in Physics* Vol. 1, January 2008. URL: www.ptep-online.com

[22] Victor Christianto. An Exact Solution of Riccati Form of Navier-Stokes Equations with Mathematica, *Prespacetime Journal* Vol. 6 Issue 7, July 2015. http://www.prespacetime.com

[23] Richard H. Enns & George C. McGuire. *Nonlinear Physics with Mathematica for Scientists and Engineers.* Berlin: Birkhäuser, 2001, p. 176-178.

[24] Sadri Hassani. *Mathematical Methods using Mathematica: For Students of Physics and Related Fields.* New York: Springer-Verlag New York, Inc., 2003.

[25] M.K. Mak & T. Harko. New further integrability cases for the Riccati equation. arXiv: 1301.5720 [math-ph]

[26] Ildus S. Nurgaliev. Cosmology without Prejudice. STFI 2014 vol. 4,URL: http://www.stfi.ru/journal/STFI_2014_04/nurgaliev.pdf; [7a] Ildus S. Nurgaliev. Singularities are averted by Vortices. *Gravitation and Cosmology* Vol. 16 No. 4 (2010) 313-315.

[27] Roustam Zalaletdinov. Averaging out Inhomogeneous Newtonian Cosmologies: II.Newtonian Cosmology and the Navier-Stokes-Poisson equations.arXiv: gr-qc/0212071 (2002)

[28] X-J. Yang, D. Baleanu, and J.A. Tenreiro Machado.Systems of Navier-Stokes equations on Cantor Sets. *Mathematical Problems in Engineering*, Vol. 2013, article ID 769724

[29] Zhao, Y., Baleanu, D., Cattani, C., Cheng, D-F., & Yang, X-J. 2013. Maxwell's equations on Cantor Sets: A Local Fractional Approach. *Advances in High Energy Physics* Vol. 2013 Article ID 686371, http://dx.doi.org/10.1155/2013/686371, or http://downloads.hindawi.com/journals/ahep/2013/686371.pdf

[30] J.D. Gibbon, A.S. Fokas, C.R. Doering. Dynamically stretched vortices as solutions of 3D Navier-Stokes equations. *Physica*D 132 (1999) 497-510

[31] Victor Christianto. A Cantorian Superfluid Vortex and the quantization of Planetary motion. *Apeiron* Vol. 11 No. 1, January 2004. URL: http://redshift.vif.com

[32] Victor Christianto. From Fractality of Quantum Mechanics to Bohr-Sommerfeld Quantization of Planetary Orbit Distance. *Prespacetime Journal* Vol. 3 No. 11 (2012). URL: http://www.prespacetime.com

[33] Victor Christianto. On Quantization of Galactic Redshift & the Source-Sink Model of Galaxies. *Prespacetime Journal* Vol. 4 No. 8 (2013). URL: http://www.prespacetime.com

[34] Siavash H. Sohrab. Turbulence and quantum mechanics from cosmos to Planck Scale. url: http://www.mech.northwestern.edu/web/people/faculty/sohrab.php

[35] B. Müller, E. Koyutürk, D. Göncü. Spatial energy distribution in a harmonic oscillator and the golden section. *INTERNATIONAL JOURNAL OF MECHANICS* Volume 10, 2016

[36] SALVATORE CAPOZZIELLO, ALESSANDRA LATTANZ. Spiral Galaxies as Enantiomers: Chirality, an Underlying Feature in Chemistry and Astrophysics. arXiv: 0509144 (2005)

[37] Lizbeth M. Fernandez-Hernandez, Mario A. Rodrıguez-Meza, and Tonatiuh Matos. Comparison between two scalar field models using rotation curves of spiral galaxies. Arxiv: 1708.06681 (2017)

[38] Mae Wan-Ho. Golden Cycles and Organic Spacetime. *The golden ratio orchestrates all of nature's cycles to create organic spacetime.*

[39] J. Klewicki et al. Self-similarity in the inertial region of wall turbulence. *PHYSICAL REVIEW* E 90, 063015 (2014)

[40] J.F. Lindner et al. Strange Nonchaotic Stars. *Phys.Rev.Lett.* 114, 054101 (2015)

[41] J.X. Mason. *Mathematics and continuing creation*. url: http://continuingcreation.org/mathematics-and-continuing-creation/

[42] Edouard B. Sonin. *Dynamics of quantised vortices in superfluids*. Cambridge: Cambridge University Press, 2016.

[43] Kerson Huang. *A superfluid Universe*. Singapore: World Scientific Publishing Co. Pte. Ltd., 2016.

[44] Sivaram, C., & K. Arun (2012) Primordial rotation of the Universe, Hydrodynamics, Vortices and angular momenta of celestial objects, *The Open Astronomy Journal*, 5, 7-11. URL: http://benthamscience.com/open/toaaj/articles/V005/7TOAAJ.pdf

[45] Victor Christianto, Florentin Smarandache, Yunita Umniyati. A Derivation of Fluidic Maxwell-Proca Equations for Electrodynamics of Superconductors & Its Implication to Chiral Cosmology Model. *Prespacetime J.,* Vol 9, No 7 (2018), url: https://prespacetime.com/index.php/pst/article/view/1471

[46] F. Smarandache & V. Christianto. Schrodinger Equation and the Quantization of Celestial Systems. *Prog. In Phys.* Vol.2, April 2006. url: http://ptep-online.com/2006/PP-05-12.PDF

[47] V. Christianto, D.L. Rapoport & F. Smarandache. Numerical Solution of Time-Dependent Gravitational Schrodinger Equation. *Prog. In Phys.* Vol.2, April 2007. url: http://ptep-online.com/2007/PP-09-11.

Opinion

The Impossibility of Direct Detection of Big Bang Relic Neutrino

Victor Christianto[1][*] & Florentin Smarandache[2]

[1]Malang Institute of Agriculture (IPM), Malang, Indonesia
[2]Dept. of Math. Sci., Univ. of New Mexico, Gallup, USA

Abstract

The existence of big bang relic neutrinos is a basic prediction of standard cosmology. It predicts the existence of 10^{87} neutrinos per flavour in the visible universe. This is an enormous abundance unrivalled by any other known form of matter except cosmic microwave background ("CMB") photons. However, unlike the CMB photons which were detected in the 1960s, the relic neutrino continue to evade direct detection due to weak interaction of neutrinos. In this paper, we argue that direct detection of relic neutrinos is impossible because of two chief reasons: (a) There was no such thing as cosmic singularity, hence the hot big bang/primeval atom model might be based on false premises; and (b) The existence of neutrino may be questionable.

Keywords: CMB, relic neutrino, direct detection, impossibility, hot big bang, cosmology.

According to standard cosmology, neutrinos should be the most abundant particles in the universe, after CMB photons. The CMB neutrino is the oldest relic, present since BBN era. However, in the past 5 decades or so, attempts to directly detect Cosmic Neutrino Background have never been succeeded.

The ideas of detecting CvB have been discussed since the 1960s. However the direct observations of the relic neutrinos is a great challenge to present experimental techniques due to the very low energy ($\sim 10^{-4}$ eV) of relic neutrinos at the present epoch [5]. It is therefore natural to ask: what are the prospects of a more direct, weak interaction based relic neutrino detection, sensitive in particular to the CvB in the present epoch. It is known, that all the existing measurements probe only the presence of the relic neutrinos at early stages in the cosmological evolution, and this often in a rather indirect way [2].

It is obvious that either WIMP or hot model of dark matter has not been observed yet. One of the most promising laboratory search, based on neutrino capture on beta decaying nuclei, may be done in future experiments designed to measure the neutrino mass through decay kinematics [3]. Another method is still underway, i.e. using PTolemy. According to Cocco [4]:

---

[*]Correspondence: Victor Christianto, Malang Institute of Agriculture (IPM), Malang, Indonesia.
Email: victorchristianto@gmail.com

ISSN: 2153-8301   Prespacetime Journal   www.prespacetime.com
Published by QuantumDream, Inc.

> The PTolemy project aim at the direct detection of the Cosmological Relic Neutrino background by the use of a Tritium target. Cosmological Relic Neutrino produced in the early stage of the Big Bang are predicted to have thermally decoupled from other forms of matter at approximately 1 second after the Big Bang; they represent the oldest detectable Big Bang relics and as such they carry an invaluable content of information about the genesis and evolution of our Universe. …In particular Tritium is among the nuclei having the most favorable detection conditions.

For a recent discussion on possible measurement of relic neutrino, see [6].

Despite all of those progress in developing measures to directly detect relic neutrino background, there is one possibility why such a direct detection remains elusive: because there was no such thing as cosmic singularity. In other words, while we accept such an initial point of creation of the Universe, its beginning came through from a *non-singular origin.*

In two recent papers, we have outlined how a non-singular origin of the Universe is possible, if we consider a turbulence model of Early Universe, because the model includes nonlinear Ermakov equation instead of Friedman equation as usual [8-9].

Taking into considerations two other findings in recent years: (a) Earth Microwave Background by P-M. Robitaille (see [10]-[13]), and (b) Theories which suggest that cosmic singularity can be removed, we submit the following hypothesis: *Direct detection of Cosmic Neutrino Background is impossible because there is no such thing as cosmic singularity.*

There are two more arguments, which seem to support our argument as outlined above: i.e. neutrino does not really exist, as well as quark matter does not exist in nature.

According to Baurov [7]:

> The analysis based on a new hypothesis that the observed physical space is formed from a finite set of *byuons*, "one-dimensional vectorial objects". It is shown in the article that the hypothesis for existence of neutrinos advanced by Pauli on the basis of an analysis of the conservation laws, is not unquestionable since the fulfillment of these laws may be secured by the physical space itself (physical vacuum) being the lowest energy state of a discrete oscillating system originating in the course of *byuon* interaction. This effect is analogous to that of Mossbauer. The direct experiments on detecting neutrinos are explained from the existence of a new information channel due to the uncertainty interval for coordinate of the four-contact *byuon* interaction forming the interior geometry of elementary particles and their properties.

Baurov also suggests that [7]:

> [A]ccording to the conception being developed on formation of physical space from a finite set of byuons, the invoking the *Pauli's hypothesis on the existence of neutrino is by no means necessary to explanation of weak interactions.*

Therefore, we argue that direct detection of cosmic neutrino background is impossible because there is no such thing as cosmic singularity.

Received April 4, 2020; Accepted May 10, 2020

References

[1] Andreas Ringwald and Yvonne Y. Y. Wong (2004). Gravitational clustering of relic neutrinos and implications for their detection. arXiv: hep-ph/0408241.

[2] Andreas Ringwald (2005). How to detect Big Bang Relic Neutrinos? DESY 05-045. arXiv: hep-ph/0505024.

[3] Andreas Ringwald (200(0. Prospects for the direct detection of the cosmic neutrino background. arXiv: 0901.1529.

[4] Alfredo G. Cocco. PTolemy - Towards Cosmological Relic Neutrino detection. *Neutrino Oscillation Workshop 4 - 11 September, 2016 - Otranto* (Lecce, Italy)

[5] Alexandre V. Ivanchik and Vlad Yu. Yurchenko (2018). Relic neutrinos: Antineutrinos of Primordial Nucleosynthesis. arXiv: 1809.03349.

[6] Stefano Gariazzo. Relic neutrinos: local clustering and consequences for direct detection. *European Physical Society Conference on High Energy Physics* - EPS-HEP2019 10-17 July, 2019 Ghent, Belgium.

[7] Yu A. Baurov (1997). Does Neutrino Really Exist? arXiv:hep-ph/9702329v1

[8] V. Christianto & F. Smarandache (2019). One note samba approach to cosmology. *Prespacetime* J., 10(6): 815-826.

[9] V. Christianto & F. Smarandache (2019). *Asia Mathematika* J., 2(2) (www.asiamath.org).

[10] Pierre-Marie L. Robitaille (2008). The Earth Microwave Background (EMB), atmospheric scattering and the generation of isotropy. *Prog. In Phys.* 2: 164-165. url: http://pteponline.com/index_files/2007/PP-10-01.PDF

[11] Pierre-Marie L. Robitaille (2009). COBE: A radiological analysis. *Prog. In Phys.* 4: p1742. url: http://ptep-online.com/index_files/2009/PP-19-03.PDF

[12] Stephen J. Crothers. COBE and WMAP: Signal analysis by fact or fiction? url: http://sjcrothers.plasmaresources.com/COBEWmap-3.pdf

[13] John Hartnett (2007). WMAP 'proof' of big bang fails normal radiological standards. *Journal of Creation* 21(2). url: http://creationontheweb.com/images/pdfs/tj/j21_2/j21_2_5-7.pdf

(Published in Prespacetime Journal | May 2020 | Volume 11 | Issue 3 | pp. 293-293)

Commentary

Unidirectional Beams

B. G. Sidharth[*]

G.P. Birla Observatory & Astronomical Research Centre,
B.M. Birla Science Centre, Adarsh Nagar, Hyderabad – 500063, India

Abstract

The unidirectional beams could be caused by a perpendicular magnetic field.

Keywords: Causality, Einstein, special relativity.

Recently it has been commented that jet streams in the cosmos or even for that matter solar wind show a departure from the laws of physics in that there are streams perpendicular to the mainstream. As shown by the author, quite some time back such an effect could be caused by a perpendicular magnetic field. This is perhaps the explanation for these cosmic jet puzzles.

The author has been studying these unidirectional beams for some time and most recently in [1]. If the velocity of the beam is $\vec{v}$ and the magnetic field is $\vec{B}$ in a perpendicular direction then the net force on the beam would be $\sim \vec{v} \times \vec{B}$ which is perpendicular to the beam. This could explain the mystery.

Received April 14, 2020; Accepted May 23, 2020

References

B.G. Sidharth (2019), Unidirectional Beams and the Virial Theorem, in *Twenty First Century Perspective of Spacetime* (Nova Science, New York).

---

*Correspondence: B. G. Sidharth, G.P. Birla Observatory & Astronomical Research Centre, B.M. Birla Science Centre, Adarsh Nagar, Hyderabad – 500063, India. Email: iiamisbgs@yahoo.co.in